Chemical Engineers in Medicine 2022

Topical Conference at the 2022 AIChE Annual Meeting

Phoenix, Arizona, USA
13-18 November 2022

ISBN: 978-1-7138-7915-2

Printed from e-media with permission by:

Curran Associates, Inc.
57 Morehouse Lane
Red Hook, NY 12571

Some format issues inherent in the e-media version may also appear in this print version.

Printed with permission by Curran Associates, Inc. (2023)

For permission requests, please contact AIChE
at the address below.

AIChE
120 Wall Street, FL 23
New York, NY 10005-4020

Phone: (800) 242-4363
Fax: (203) 775-5177

www.aiche.org

Additional copies of this publication are available from:

Curran Associates, Inc.
57 Morehouse Lane
Red Hook, NY 12571 USA
Phone: 845-758-0400
Fax: 845-758-2633
Email: curran@proceedings.com
Web: www.proceedings.com

TABLE OF CONTENTS

CHEMICAL ENGINEERING PRINCIPLES ADVANCING MEDICINE II

153e Genetically Stable CRISPR-Based Kill Switches for Engineered Microbes 1
Austin Rottinghaus, Aura Ferreiro, Gautam Dantas, Tae Seok Moon

219b Virus Encapsulation in Polypeptide Complex Coacervates for Vaccine Formulations 3
Caryn Heldt, Pratik Joshi, Claire Decker, Xianci Zeng, Arvind Sathyavageeswaran, Sarah Perry

219c Association and Adsorption of Mucins to the Air Interface in Human Airway Mucus 4
Scott Danielsen, Richard C. Boucher, Michael Rubinstein

219d Computational Fluid Dynamics to Understand Ureteroscopy Irrigation 5
David G. Foster

219e Validation of Stroke Prediction in Patients with Carotid Artery Disease Using CFD 7
David G. Foster, Lauren Redus

219f A Computational Predictive Framework Towards Individualized Risk Assessment of Kidney Transplantation Failure 9
Symeon Savvopoulos, Andreas I. Reppas, Wilfried Gwinner, Irina Scheffner, Haralampos Hatzikirou

219g Monte Carlo Simulations of Spherocylinders Interacting with Site-Dependent Square Well Potentials 11
Kiranmai Yellam, Prateek Jha

MEDICAL DEVICES

89a Low-Level Metal Toxicology in Serum Via Anodic Stripping Voltammetry 12
Shaylee R. Larson, Mary Jeppson, Kevin Chandler, Paige Leland, Himanshu Sant, Swomitra Mohanty

89b Comparison of Methemoglobin and Deoxyhemoglobin As Potential Contrast Agents in Magnetic Resonance Imaging (MRI) 14
Royaalsadat Ayati, Kyle Manwaring, Randy S. Lewis

89c Barrier-Free Paper Analytical Devices for Multiplex Colorimetric Detection 15
Ayushi Chauhan, Bhushan Toley

89d Clinical Validation of Breath Biomarkers in Pediatric Tuberculosis Patients Using Comprehensive Two-Dimensional Gas Chromatography Time-of-Flight Spectrometry 17
Mary Jeppson, Emily Lym, Swomitra Mohanty

89e Elucidating the Effects of Shear and Surface Topography on Thromboembolism in Ventricular Assist Devices 19
Anjana Jayaraman, Junhyuk Kang, James Antaki, Brian J. Kirby

89f Active Levy Swimmers and Geometric Design for Anti-Infection Catheters 22
Tingtao Zhou

89g A Low Cost, Compact Flow Cytometer 23
Mahrukh Mir, Mahesh S. Tirumkudulu

CHEMICAL ENGINEERING PRINCIPLES ADVANCING MEDICINE I

153a Slow-Release Nanoparticles for Chronic Pain Management 24
Parker Lewis, Rachel Pollard, Rocco Latorre, Dane Jensen, Brian Schmidt, Nigel Bunnett, Nathalie M. Pinkerton

153b Pseudomonas Aeruginosa Reverse Diauxie Is a Multidimensional, Optimized, Resource Utilization Strategy Facilitating Chronic Wound Colonization 25
S. Lee McGill, Yeni Yung, Kristopher A. Hunt, Michael Henson, Luke Hanley, Ross P. Carlson

153d Evaluating Antimicrobial Drug Formulations on a 3D Biofilm Growth Model Under Simulated In Vivo Conditions 26
Yadiel Varela-Soler, Himani Chavda, Swarnima Roychowdhury, Veronica Farag, Patrick Erickson, Biju Parekkadan, Charles Roth

153f Improving Therapeutic Protein Secretion in Probiotic Yeast Saccharomyces Boulardii using a Multifactorial Engineering Approach. 27
Deniz Durmusoglu, Ibrahim Al'Abri, José Luis Martinez Ruiz, Nathan Crook

153g A Low-Cost Strategy for Genotypic Antimicrobial Resistance Detection Using Oligonucleotide Ligation Assay 28
Ayushi Chauhan, Bhushan Toley

ENGINEERING CANCER I: MECHANISTIC STUDIES

526a The Effects of Intratumoral Heterogeneity on Metastasis of Triple-Negative Breast Cancer Cells 30
Molly Brennan, Susan E. Leggett, Sophia Martinez, Celeste M. Nelson

526b The Effects of Palmitate-Induced IRE1α Activation in DNA Double Strand Break Repair and the Development of Chemotolerant Breast Cancer Cells 31
Kevin Chen, S. Patrick Walton, Christina Chan

526c Macrophages Aggregate and Cooperate to Phagocytose Cohesive Tumor Cell Targets in Engineered Immuno-Tumoroids and Solid Tumors 32
Lawrence J. Dooling, Jason C. Andrechak, Brandon H. Hayes, Siddhant Kadu, Dennis E. Discher

526d Mechanistic Characterization of Homologous Recombination in BRCA2 Mutant Cancers 33
Shayne Sensenbach, Prashant Karki, Mehmet Orman

526e Modeling Cancer Dormancy and Recurrence with the Theory of Birth-Death Processes 34
Adeyinka Lesi, David Rumschitzki

526g Machine-Learning Based Prediction of Membranolytic Peptides with Anticancer Activities 35
Atefe Alimirzaei, Christopher Kieslich, Hyeju Song

526h Caspase-Dependent HDAC4 Translocation Due to Microsecond Pulsed Electric Field (μsPEF) Exposure of Glioblastoma Cells 36
Zahra Safaei, Gary Thompson

TISSUE ENGINEERING, BIOPRINTING, AND REGENERATIVE MEDICINE

471a Mitochondrial Mechanisms Underlying NANOG Induced Reversal of Aging 39
Debanik Choudhury, Na Rong, Nika Rajabian, Georgios Tseropoulos, Pihu Mehrotra, Ramkumar Thiyagarajan, Pedro Lei, Izuagie Ikhapoh, Kenneth Seldeen, Bruce Troen, Stelios Andreadis

471b Effects of Processing Conditions on the Macroscopic Properties of Cellulose Filled Hydrogel Scaffolds Using UV Rheology 40
Bobby Haney, Subramanian Ramakrishnan

471c 3D Bioprinting of iPSC Derived Islet Organoids in Hydrogel Constructs 41
Miranda Poklar, Ben Mizerak, Ravi Krishnamurthy, Connor Wiegand, Prashant Kumta, Ipsita Banerjee

471d Engineering the Skeletal Muscle for Improved Innervation after Peripheral Nerve Injury 43
Pihu Mehrotra, Shahryar Shahini, Nika Rajabian, Yali Zhang, Jianmin Wang, Song Liu, James Jablonski, Susan Udin, Stelios Andreadis, Kirkwood Personius

471e Regeneration of the Epithelium Layer Using Fibrous Biomimetic Basement Membranes in a Mouse Dermal Wound Model 44
Dina Gadalla, Maeve Kennedy, David Lott

471f Photochemical Reaction Kinetics and Thermodynamics of Light-Induced Collagen Cross-Linking with Rose Bengal for Suture-Less Wound and Incision Closure and Repair 46
Alan Aguirre-Soto

BIG DATA AND MACHINE LEARNING TO ADVANCE MEDICINE

284a Data-Driven QSAR Modeling for the Iterative Identification of Chemical Motifs from Limited Data 47
Hyun-Seob Song, Chul Min Park, Soojin Jang, Jae Sun Kim, Hyung Chul Ryu, Hasoo Seong

284b Single Sequence Prediction of Protein Structure and Impacts on Computational Protein Design 48
Ratul Chowdhury

284c Hierarchical Graph-Based Representation Drives Prediction of Stapled Peptide Drug-like Properties 49
Marshall Case, Camille Bilodeau, Greg Thurber

284d Data-Driven Approach for the Prediction of MHC Class II Epitopes Using Oscillations of Physicochemical Properties 52
Hyeju Song, Christopher Kieslich

284e Automated Detection of Apoptotic Bodies in Label-Free Time-Lapse High-Throughput Microscopy Using Deep Convolutional Neural Networks 53
Kwan-Ling Wu, Melisa Martinez-Paniagua, Kate Reichel, Prashant Menon, Shravani Deo, Badrinath Roysam, Navin Varadarajan

284f Single-Cell Transcriptomic Analysis of Neural Development 54
Karthik Shekhar

284g Structure-Guided Functional Phylogenetic Trees of Human Channel Proteins and Kinases for Drug Discovery 55
Ratul Chowdhury

284h Machine Learning Enabled Cancer and Immune Cells Segmentation on Clinical Images 56
Vidit Shah, Felicia Poynter, Shachi Mittal

PANDEMIC RESPONSE AND PUBLIC HEALTH

348a Indoor Transport of Aerosolized Respiratory Droplets in Multiroom Buildings 58
Alex Vlachokostas, Timothy Salsbury, Carolyn A. Burns, Leonard Pease

348b Multi-Strain Integrated Modelling for COVID-19 59
Dimosthenis Sarigiannis, Achilleas Karakoltzidis, Spyros Karakitsios

348c Portable and Label-Free Quantitative Loop-Mediated Isothermal Amplification (LF-qLamp) for Reliable COVID-19 Diagnostics in Three Minutes of Reaction Time 62
Sergio Bravo-González, Grissel Trujillo De Santiago, Mario Moisés Álvarez

348d Affordable Rapid PCR-Based Biodetection at the Point of Care Via Rayleigh-Bénard Convection 63
Mingin Kim, Victor Ugaz

348f Rapid Bioproduction of Protein Biologics Using Plants in Response to Sars-Cov-2 64
Katherine Haddad, Seongwon Jung, Imran Khan, Nancy E. Lane, Somen Nandi, Karen A. McDonald

348g Strategic Planning of Joint COVID-19 Booster and Influenza Vaccination Campaign: The UK COVID-19 Pandemic Study 66
Dauda Ibrahim, Zoltán Kis, Maria Papathanasiou, Cleo Kontoravdi, Benoit Chachuat, Nilay Shah, Miriam Sarkis

ENGINEERING CANCER II: THERAPY

573a Low Doses of Paclitaxel Induce a Dormant State in Brain Metastatic Breast Cancer Spheroids 68
Raghu Vamsi Kondapaneni, Rachel Warren, Shreyas Rao

573b Engineer CAR-Neutrophils for Targeted Cancer Immunotherapy 69
Xiaoping Bao

573c Immunogenic Treatment for Metastatic Breast Cancer Using Targeted Carbon Nanotube Mediated Photothermal Therapy in Combination with Anti-PD-1 70
Gabriela Faria, Alexis Woodward, Clément G. Karch, Adam Aissanou, Roger Harrison

573d Immune Cell Homing Materials for Cancer Immunotherapy 72
Hua Wang

573e Electroporation and Cold Atmospheric Plasma As a Novel Cancer Treatment 73
Jordan Hoops, Kristen I. Haller, Mikaya M. Elliott, Rylie N. Andrews, Nicole Miller, Timothy Brenza, Prasoon K. Diwakar

573f Combined Gene and Chemotherapy for Cancer: Interdisciplinary Multimodal Approaches to Effective, Safe, and Diverse Therapies 74
Rebecca Lee, Cheng Wai "winne" Lei, Margaret Lugin, Jee Young Chung, Woo Chang Hwang, Angela G. Fleischman, Namshik Han, Young Jik Kwon

573g The Landscape of Drug Sensitivity in Cancer Cell Lines Reveals Effective Drug Combinations for Cancer 75
Belinda Garana, Nicholas Graham

573h Multi-Modal Single-Cell Characterization As a Basis for Precision Therapy for Intratumoral Heterogeneity in Glioblastoma 76
James Park, Abdullah Feroze, Samuel Emerson, Anca Mihalas, C. Keene, Patrick Cimino, Adrian Lopez Garcia De Lomana, Wei-Ju Wu, Serdar Turkarslan, Kavya Kannan, Nitin Baliga, Anoop Patel

INFECTION AND PREVENTION, EPIDEMIOLOGY AND TREATMENTS, DIAGNOSTIC APPROACHES

417a Two Complementary Memory B Cell Processes Generate Antibody Protection Against the Same or Variant Viruses 77
Matthew Vanbeek

417b A Model for How T Cell-Mediated Autoimmunity Can be Triggered By Persistent Viral Infections 78
Rose Yin, Samuel Melton, Eric Huseby, Mehran Kardar, Arup K. Chakraborty

417c Predicting the Effects of Sars-Cov-2 Spike Protein Mutations to MHC Class II – Mediated Immune Responses 79
Mercedes Haley, Sumaiya Islam, Robert Pantazes

417d Inhibition of Sars-Cov-2 Spike Protein Pseudotyped Virus Infection Using ACE2-Tethered Micro/Nanoparticles 80
Soha Alkhaldi, Ian Peng, Esmael Alyami, Ammar Ahmad Tarar, Ching-An Peng

417e A Novel SERS-Based Assay for Detection of the Sars-Cov-2 Spike Protein 81
Karen Wawrousek, Moein Mohammadi, Delphine Antoine, Madison Vitt, Sharmin Sultana Jyoti, Julia Dickie, Gerard Wall, Patrick A. Johnson

417f Cell-Free, Dendritic Cell-Mimicking Vaccines for Cancer, COVID-19, and Beyond 82
Melissa N. Thone, Jee Young Chung, Young Jik Kwon

417g Agent-Based Modeling of Sars-Cov-2 Transmission during the COVID-19 Pandemic at the University of Delaware 83
Soham Jariwala, Norman J. Wagner, Antony Beris, Richard Suminski, Gregory Dominick

ENGINEERING CANCER III: DEVICES FOR DIAGNOSIS, CULTURING, AND MICROENVIRONMENT STUDIES

616a Nature-Inspired 3D Scaffolds to Improve Ex Vivo T-Cell Culturing Environments for Adoptive Cell Transfer Cancer Immunotherapy 85
Lucy Todd, Matthew Chin, Marc-Olivier Coppens

616b Isolation and Identification of Subpopulations of Circulating Tumor Cells Using a Microchip with Bio-Inspired Patterns 86
Wei Li, Zhenya Ding

616c Dynamic 3D Co-Culture of Stromal and Immune Cells with ER^+ Breast Cancer Using a Thiol-Acrylate Hydrogel Scaffold and Microfluidic Droplet Trapping Array 88
Braulio Ortega Quesada, Anowar H. Khan, Sophia Zhou, Elizabeth C. Martin, John Pojman, Adam Melvin

616d Cell-Free Protein Synthesis Biosensing of Glutamine in Human Serum and Saliva: Towards at-Home Low-Cost Diagnostics for Personalized Cancer Treatment 89
Brad Bundy, J Porter Hunt, Mehran Soltani

616e Investigating Intracellular Adhesion and Intercellular Signaling in Glioblastoma Models 90
Rosalyn Hatlen, Padmavathy Rajagopalan

616f Probing the Spatiotemporal Attributes of Gamma-Secretase/Notch Signaling in Breast Cancer Microenvironments.. 92
Malcolm Lane Gilchrist, Yueming Li

616g Multivariate Model of an Engineered Niche Delineates Metastatic Potential of Breast Cancer 93
Sophia Orbach, Christian Devaull, Elizabeth Bealer, Jacqueline Jeruss, Lonnie Shea

616h Multimodal Imaging for Simultaneous Measurement of Melanocyte Mass and Pigmentation................ 96
Tarek Moustafa, Edward Polanco, Rachel Belote, Robert Judson-Torres, Thomas Zangle

Author Index

153e Genetically Stable CRISPR-Based Kill Switches for Engineered Microbes

Austin Rottinghaus, Aura Ferreiro, Gautam Dantas, Tae Seok Moon

Microbial biocontainment is an essential goal for engineering safe, next-generation living therapeutics [1, 2]. However, the genetic stability of biocontainment circuits, including kill switches, is a challenge that must be addressed. Kill switches are among the most difficult circuits to maintain due to the strong selection pressure they impart, leading to high potential for evolution of escape mutant populations. We engineered two CRISPR-based kill switches in the probiotic Escherichia coli Nissle 1917, a single-input chemical-responsive switch and a 2-input chemical- and temperature-responsive switch. We employed parallel strategies to address kill switch stability, including functional redundancy within the circuit, modulation of the SOS response, antibiotic-independent plasmid maintenance, and provision of intra-niche competition by a closely related strain. We demonstrate that strains harboring either kill switch can be selectively and efficiently killed inside the murine gut, while strains harboring the 2-input switch are additionally killed upon excretion. Leveraging redundant strategies, we demonstrate robust biocontainment of our kill switch strains and provide a template for future kill switch development [3]. In this talk, we will discuss our recent efforts to expand this technology into engineering diverse bacteria, including cyanobacteria, soil bacteria, and gut and soil microbial consortium, although the focus is to engineer the gut microbiota.

1. ***Amrofell, Rottinghaus, and Moon, Curr Opin Biotechnol, 2020. 66: p. 11-17.***
2. ***Rottinghaus, Amrofell, and Moon, Biotechnology Journal, 2020. 15: p. 1900319.***
3. ***Rottinghaus et al. Nature Communications, 2022. 13, Article number: 672***

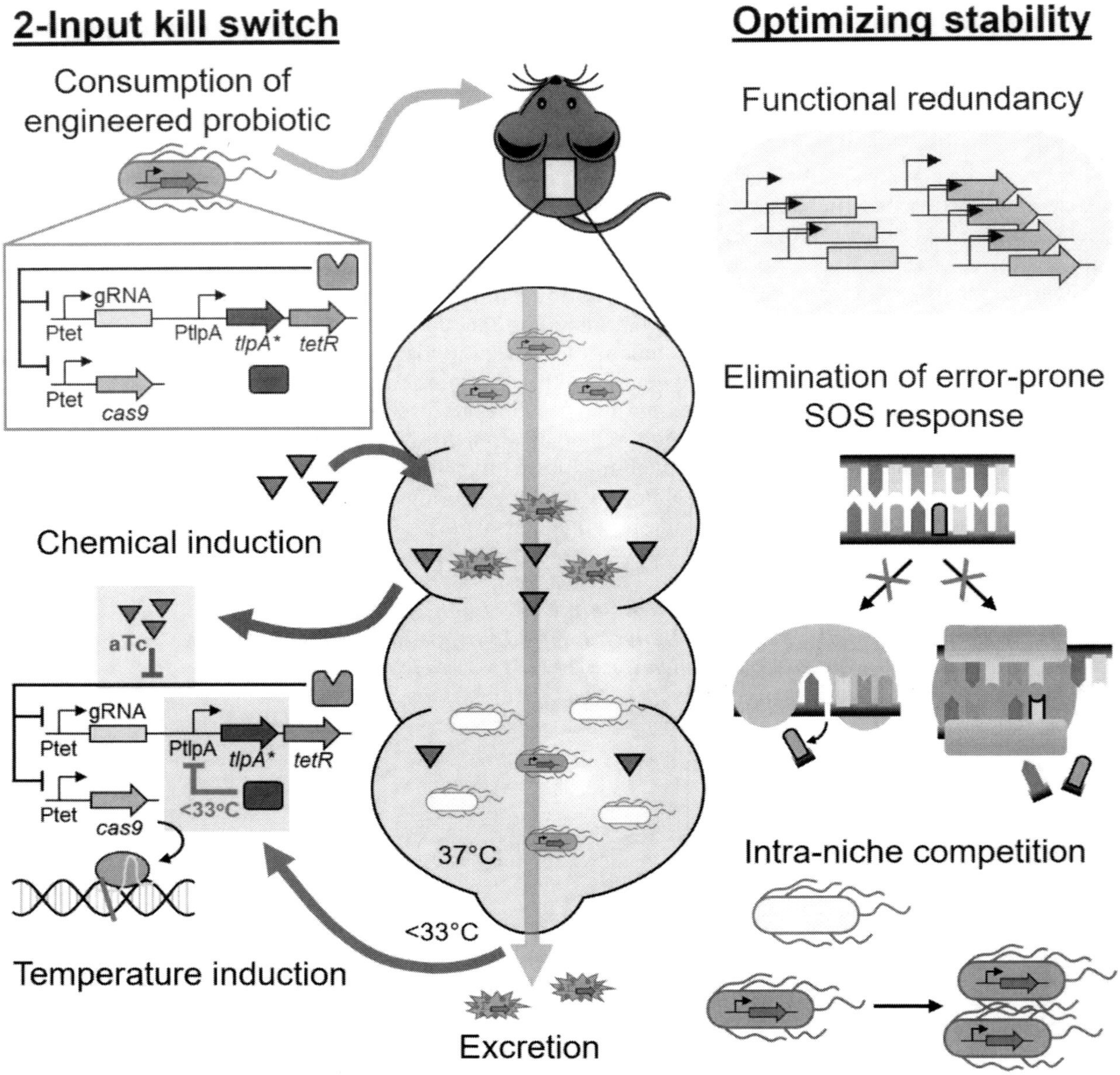

2-Input kill switch
Consumption of engineered probiotic
gRNA
Ptet
PtlpA
tlpA*
tetR
Ptet
cas9
Chemical induction
aTc
gRNA
Ptet
PtlpA
tlpA*
tetR
<33°C
Ptet
cas9
Temperature induction
37°C
<33°C
Excretion
Optimizing stability
Functional redundancy
Elimination of error-prone SOS response
Intra-niche competition

219b Virus Encapsulation in Polypeptide Complex Coacervates for Vaccine Formulations

Caryn Heldt, Pratik Joshi, Claire Decker, Xianci Zeng, Arvind Sathyavageeswaran, Sarah Perry

The lack of shelf stability of biological therapeutics can cause large disruptions in distribution of life saving medicines. We have been exploring the ability of polypeptide complex coacervates to stabilize viral vaccine models at higher temperatures. This would increase the equitable distribution of vaccines worldwide. Charge patterning of peptide coacervates and the charge patterning on the virus surface highly influence encapsulation and thermal stability of virus. Here we show that porcine parvovirus (PPV), which is highly charged, has a high encapsulation efficiency and it can be influenced by peptide charge density during coacervation. However, human rhinovirus 14 (HRV) has a lower charge on its surface and encapsulation into the coacervate phase is halted when charge patterning is introduced. Coacervate stability can be altered by changes in the cationic polymer and hydrophobicity of the polymer. The general encapsulation trends are different for viruses than for proteins. This demonstrates the need to continue to explore methods to stabilize viral vaccine formulations and polypeptide complex coacervates may hold important keys to understanding the stability of future viral vaccine formulations.

219c Association and Adsorption of Mucins to the Air Interface in Human Airway Mucus

Scott Danielsen, PhD, Richard C. Boucher, Michael Rubinstein

The airway surface layer lines the respiratory tract, simultaneously trapping inhaled particulates and facilitating their removal from the lung. Secreted gel-forming mucins, which possess both hydrophilic glycosylated domains and hydrophobic globular domains, interact with a wide range of proteins and nucleic acids and are responsible for the characteristic biophysical properties of mucus. The mucin hydrophobic domains promote: 1) self-association in the bulk phase; and 2) strong adsorption at the air-mucus interface with formation of a thin, viscoelastic skin layer that effectively separates the mucus layer itself into two distinct sublayers. Measuring the modulus of mucus using multiple rheometric geometries with varying surface-to-volume ratios, we measure different apparent viscosities with different contributions of surface to bulk. A simple model deconvolutes these contributions, permitting assessment of the bulk and surface properties. Surfactants are shown to reduce the associative interactions of mucins, lowering the degree of association/adsorption and overall viscoelastic modulus. Understanding of the structural organization of mucus in the airway and resultant mechanical properties will enable the development of therapeutic approaches to improve mucus clearance in muco-obstructive lung diseases.

219d Computational Fluid Dynamics to Understand Ureteroscopy Irrigation

David G. Foster

Ureteroscopy with laser lithotripsy is a minimally invasive kidney stone treatment that is a common process by urologists and is becoming more important with the lifetime prevalence of kidney disease. [1] The treatment is primarily used for kidney stone removal when stones cannot be passed by the patient. The procedure involves the insertion of an ureteroscope, a very narrow scope, that allows urologists to travel up the ureter to the kidney that includes a camera to assist in finding the stones and monitoring the removal process. Ureteroscopy uses the energy pulse of a laser beam, applied directly to the stone, to chip it away into pieces small enough to pass through the ureter. The treatment also involves a process called ureteroscopy irrigation, in which a pressurized channel of saline solution is injected into the ureter and the kidney. Ureteroscopy irrigation is used to secure a better field of vision inside the kidney, it also helps provide a spontaneous flow passage for kidney stone fragments to move out of the kidney and into the bladder without using any additional instrumentation.[2] As a result, the irrigation aids urologists in locating the precise location of the stones in the kidney and helps decrease procedure time. The goal of this project is for the first time to compare and contrast laboratory measurements of the ureteroscopy irrigation process for flow velocity and pressure with simulated values obtained from computational fluid dynamics (CFD) simulations.

CFD is a branch of mechanics that creates simulations for a variety of applications and is becoming a useful tool in medical applications to study the physiological flow patterns of fluids in the human body.[3] We have recently found that CFD can accurately predict the flow patterns and streamlines of flow in the kidney and has correctly predicted where new stones will form based on the flow obstruction patterns when other stones are present. These findings have led to the construction of an in vitro hydrogel model of an actual patient kidney where measurements of velocity and pressure can be made at different locations and compared to the CFD simulations. Geometries were obtained from scans of patients' anatomy and inserted into the CFD program. The ureteroscope was drawn into the geometry and placed at 5 different locations within the geometry to obtain representative fluid velocity and intrapelvic pressure values that were validated with measured values from the physical model. The main challenge of this work is to create an accurate 3-dimensional mesh throughout the geometry where the CFD software uses finite volume methods to solve the basic equations of fluid dynamics using iterative procedures to obtain several million data points throughout the geometry. The optimization and manipulation of the geometries to create an accurate mesh is generally the rate limiting step within the CFD simulation. The quality of the mesh corresponds directly to the accuracy of the simulation. The scope placement must be realistic and accurate enough for the simulation to be representative of the procedure.

This project involves the application of CFD modeling of the fluid velocity and intrapelvic pressure during ureteroscopy. The results have provided a novel way to computationally simulate a patient's specific and unique flow patterns and pressure gradients during ureteroscopy procedures. Additionally, CFD can be used to establish a heat transfer profile during laser lithotripsy. The first goal of the project is to compare the physical and CFD models for velocity and intrapelvic pressure profiles corresponding to different irrigation velocities and pressures. Using CFD simulations extensive velocity and pressure profiles were generated within the kidney and ureter, providing optimal settings to the urologist. The second goal of the project is to examine the heat transfer mechanism of the laser lithotripsy. Optimal irrigation velocities and pressures have been a topic of debate among urologists due to being solely based on empirical results. Using CFD we are able to obtain time-dependent profiles of the heat transfer process in order to establish a safe working range for the laser lithotripsy process. These simulations could significantly advance the confidence that this procedure is being done as safely as possible.

References:

1. University of Wisconsin Hospitals and Clinics Authority. "How Common Are Kidney Stones?" *UW Health*, uwhealth.org/urology/how-common-are-kidney-stones/11208.
2. Wright, Anna E et al. "Ureteroscopy and stones: Current status and future expectations." *World journal of nephrology* 3,4 (2014): 243-8. doi:10.5527/wjn.v3.i4.243
3. Williams, Jessica G et al. "The Fluid Mechanics of Ureteroscope Irrigation." *Journal of endourology* 33,1 (2019): 28-34. doi:10.1089/end.2018.0707

219e Validation of Stroke Prediction in Patients with Carotid Artery Disease Using CFD

David G. Foster, Lauren Redus

Stroke remains one of the leading causes of death in the United States accounting for one out of every twenty deaths.[1] Carotid artery stenosis (CAS) is a major cause of both stoke and transient ischemic attacks (TIAs), and it has been documented that the risk of development of these illnesses are directly related to the degree of stenosis.[2] Research has proven that carotid endarterectomies (CE) can be beneficial in treating symptomatic patients with high-grade internal carotid artery (ICA) stenosis. However, treatment of lower risk and asymptomatic patients remains controversial due to high risks associated with the procedure. Some of these risks include myocardial infarction, cranial nerve injury, hemodynamic instability, infection, and bleeding.[3] Therefore, minimally, or non-invasive procedures are currently being studied to assess patients and reduce risks of thromboembolic and ischemic events. The carotid artery carries oxygenated blood from the heart to the head. There are two carotid arteries on each side of the neck. Each common carotid artery branches into two divisions: the internal carotid, supplying blood to the brain, and the external carotid that provides blood to the face. As the carotid ascends towards the head, it branches out and creates a bifurcation. The internal carotid is especially interesting as it contains a bulge, the carotid sinus, with a diameter greater than that of the common carotid.[4,5] This geometry and the resulting increase in adverse pressure gradient create a flow reversal region during the cardiac cycle. Velocity profiles in the carotid sinus may be linked to stroke, as this site is prone to plaque deposition.[6,7] The goal of this project is to investigate velocity profiles and pressure gradients in the carotid artery to understand the biological significance of the carotid sinus and its relation to stroke risk.

Using Computational Fluid Dynamics (CFD) simulations of the carotid artery, medicine and engineering intersect. CFD uses finite-element analysis to determine flow characteristics while paying specific attention to the geometry, velocity profiles, streamlines, and pressure gradients. This method has an advantage over solely relying on ultrasound data because it provides insight into the unique flow patterns in a patient's individual carotid artery geometry. Once a mesh, a collection of vertices, edges, and faces that define the shape of an object in computational simulations, has been created, physiological boundary conditions are defined.[6,7] These boundary conditions include important parameters for blood density, blood viscosity (Carreau model), outlet pressures, body temperature, and peak systolic (PSV) and end diastolic (EDV) velocities. The Navier-Stokes Equations are solved simultaneously for flow simulations at nodes created in the meshing portion of the application. A solution is generated when residual limits are met.

This project consists of two Phases. In Phase 1, forty-five patients whose outcomes were known to the analysis team were provided by the Department of Neurosurgery at the University of Rochester Medical Center, along with their respective ultrasound data. These patient data were analyzed using CFD to determine the parameters that can be used to predict stroke. In this study, it was found that high pressure and velocity gradients near the wall of the internal carotid artery were clear indicators of the potential for stroke, and have been previously presented.[8] Now, we present the results of Phase 2 of the project where an additional forty-five patients were analyzed in a double-blind study where the patient outcomes regarding stroke were not known until after the analysis was completed. The results suggest that CFD can be used to accurately predict the outcome of stroke or no-stroke for patients with carotid artery disease. The results were consistent with our previous findings that stroke patients have irregular velocity gradients and more extreme velocities in the carotid sinus, where there is less fluid due to plaque deposit from atherosclerosis. It was also found that comparisons of patients with different stroke outcomes require more than a comparison of velocities and geometries in a steady model. A pulsatile flow user-defined function (UDF) for a transient set-up has been developed to consider velocity differences during diastole and systole. In theory, a pulsatile flow model simulates physiological conditions more realistically, as the heart pumps blood in a pulsatile fashion. Validation of the model confirmed that mass conservation is obeyed. A scaled ratio, in which the maximum velocity obtained in the CFD measurements at the peak of the velocity profile at the carotid bulge was compared to the lowest velocity at the reversal region, was also created. Preliminary comparison has shown that stroke patients have higher ratios of flow reversal. A UDF that instantly extracts velocity and pressure data at the bulge is being developed to streamline the process of correlating patient stroke outcomes. Finally, improvements to the model include accounting for the fluid-structure interaction between the artery and the blood, in which the vessel wall is modeled as a homogenous, isotropic, viscoelastic material. Ultimately, this project creates a prediction model for stroke clinical outcomes, with preventative medicine benefits.

References

[1] 'Stroke | cdc.gov', Centers for Disease Control and Prevention, 2020. [Online]. Available: https://www.cdc.gov/stroke/index.htm.

[2] 'Types of Stroke | cdc.gov', Cdc.gov, 2020. [Online]. Available: https://www.cdc.gov/stroke/types_of_stroke.htm#ischemic.

[3] A. Greenstein et al., 'Association between minor and major surgical complications after carotid endarterectomy: Results of the New York Carotid Artery Surgery study', Journal of Vascular Surgery, vol. 46, no. 6, pp. 1138-1146, 2007.

[4] Bluestein D. Utilizing Computational Fluid Dynamics in Cardiovascular Engineering and Medicine—What You Need to Know. Its Translation to the Clinic/Bedside. Artificial organs. 2017;41:117-121.

[5] J. Krejza et al., 'Carotid Artery Diameter in Men and Women and the Relation to Body and Neck Size', Stroke, vol. 37, no. 4, pp. 1103-1105, 2006.

[6] Foster, D.; Salerno, D. "A Computational Fluid Dynamics Model of the Carotid Artery". Presented at the 2017 AIChE Annual Meeting, Minneapolis, MN, October 29-November 3, 2017.

[7] Foster, D.; Weldy, A. "Computational Fluid Dynamics Simulations of the Human Carotid Artery to Predict Strokes in Patients with Carotid Artery Disease". Presented at the 2019 AIChE Annual Meeting, Chemical Engineering Principles Advancing Medicine Session, Orlando, FL, November 12, 2019.

[8] Maruicio Araiza Canizales, Priscila Passerotti Vaciski Barbosa, Lauren E. Redus, Leonor N. Teles, Hon Sum A. Lee, Jonathan Stone, and David G. Foster, "Patient-Specific CFD Simulations of the Carotid Artery to Predict Stroke," 2021 AICHE Annual Meeting, Boston, MA, November 2021.

219f A Computational Predictive Framework Towards Individualized Risk Assessment of Kidney Transplantation Failure.

Symeon Savvopoulos, Andreas I. Reppas, Wilfried Gwinner, Irina Scheffner, Haralampos Hatzikirou

Kidney transplantation is the treatment needed for end-stage renal disease, whose success depends on several factors such as cancer, serious infections, severe cardiovascular diseases, and liver disease [1,2]. After receiving a kidney transplant, patients must take immune-suppressing medications to help their immune system to accept the new organ. More precisely, after kidney transplantation, patients might experience either B, or T, or both B and T cell rejection. Among them, T cell rejection is quite common; both B and T cells rejections are less often, while B cell rejections are rare and occur after a long time. Appropriate therapeutic intervention may successfully cure these complications but in many cases, only leads to partial or no improvement. Hence, despite all previous diagnostic and therapeutic progress, the mean kidney graft survival is limited to approximately 15 years [3,4].

Glomerular filtration is the process by which the kidneys filter the blood, removing excess wastes and fluids, and indicates kidney function. The physicians estimate the Glomerular filtration rate (GFR) based on the blood creatinine, age, body size and gender. The GFR levels help a doctor to plan treatment for the patient. The lower the GFR number is, the worse kidney function is. Thus, when the diagnosis becomes early, the kidney failure progress could be decreased or be stopped [5]. Chronic kidney disease is diagnosed when the GFR is below 60 for three months or more or GFR is above 60 with kidney damage (high levels of albumin patients' urine). There are five kidney disease stages. In stage 1, the GFR is 90 or higher, and the kidney function is normal. The GFR in stage 2 ranges from 60 to 89, and there is a mild loss of kidney function. Stage 3 is divided into Stage 3a which is indicated by GFR values ranging from 45 to 59 with mild to moderate kidney function, and Stage 3b where the GFR reaches 30, with moderate to severe kidney duction. In stage 4, the loss in kidney function is severe and its GFR values are within 15 to 29. Lastly, in stage 5, the GFR drops from 15, and kidney failure appears [6].

Management of patients with chronic kidney diseases with or without transplantation is a very challenging task. During chronic diseases, there is a dynamic disease process of injury and repair responses inherent to the specific disease present. These occurring responses may be simulated using a set of interacting and disease-modifying variables, whose values are based on the conditions, the events, and potential interventions. Validating the variable complex interplay is the key to personalized patient treatment in terms of diagnostic measures and therapeutic decisions. Of course, it is required the identification of the relevant causative and non-causative risk factors, their potential interaction, estimation of putative therapeutic effects concerning the clinical disease course and mid- and long-term outcomes [7].

Common mathematical approaches for predicting GFR are regression analysis and Cox proportional hazard analysis, which numerically estimate the outcome of GFR in a given study population at a specific point in time [8]. By nature, these approaches cannot truly capture the dynamic disease processes. Also, appropriate appreciation of potential interactions between different factors in the model is challenging. Thus, such mathematical models differ fundamentally from the holistic perspective of the clinician, who tries to integrate all information about the disease course and possible relationships between various factors up to the current time of patient assessment.

By using sparse medical data, this work proposes a novel framework that can predict the dynamics of the GFR with a mathematical model and apply uncertainty analysis to predict the risk of kidney failure up to two years. The mathematical model used was similar to a dumpling oscillator with a stability threshold and a dumpling factor. Two patient datasets were used. The training set with the validation cohort consists of 1700 patients. Briefly, the initial steps of our methodology were the feature selection in the patients' unmodeled data, the regression of GFR in a year, and the parameter estimation in non-rejection cases. The model analysis found the personalized stability threshold whose values were within the literature range. Based on the final personalized GFR range values, model parameter values were estimated, and uncertainty analysis was performed with the distribution of the dumpling factor. This procedure was applied for patients without a rejection, with one rejection, and with multiple rejections. By comparing the simulated values of the uncertainty analysis with the actual data, and in different personalized threshold ranges, the methodology achieved an accurate prediction of the success of transplantation, with an area under the curve (AUC) close to 0.80.

This work shows that the combination of machine learning, mathematical modeling, and big data analysis are undoubtedly valuable for predicting kidney transplantation outcomes. Subsequently, by capturing the personalized GFR dynamics, the medical community will proceed efficiently to the personalized treatment of the patients.

1. National Institute of Diabetes and Digestive and Kidney Diseases [U.S.]. Division of Kidney Urologic and Hematologic Diseases., Urban Institute. Renal Research Program., University of Michigan., USRDS Coordinating Center. United States Renal Data System. Annual Data Report. 2008;v.
2. Humar A, Matas AJ. Surgical complications after kidney transplantation. Semin Dial. 2005;18(6):505–10.
3. Luque S, Lúcia M, Melilli E, Lefaucheur C, Crespo M, Loupy A, et al. Value of monitoring circulating donor-reactive memory B cells to characterize antibody-mediated rejection after kidney transplantation. Am J Transplant. 2019;19(2):368–80.
4. Matz M, Fabritius K, Lorkowski C, Dürr M, Gaedeke J, Durek P, et al. Identification of T cell-mediated vascular rejection after kidney transplantation by the combined measurement of 5 specific MicroRNAs in blood. Transplantation. 2016;100(4):898–907.
5. Bakris GL, Williams M, Dworkin L, Elliott WJ, Epstein M, Toto R, et al. Preserving renal function in adults with hypertension and diabetes: a consensus approach. National Kidney Foundation Hypertension and Diabetes Executive Committees Working Group. Am J Kidney Dis [Internet]. 2000;36(3):646–61. Available from: http://www.ncbi.nlm.nih.gov/pubmed/10977801
6. Thomas C, Thomas L. Renal failure--measuring the glomerular filtration rate. Dtsch Arztebl Int (Internet). 2009;106(51–52):849–54. Available from: http://www.pubmedcentral.nih.gov/articlerender.fcgi?artid=2803612andamp;tool=pmcentrezandamp;rendertype=abstract
7. Ng K, Kartoun U, Stavropoulos H, Zambrano JA, Tang PC. Personalized treatment options for chronic diseases using precision cohort analytics. Sci Rep. 2021;11(1).
8. Van Den Brand JAJG, Dijkstra TMH, Wetzels J, Stengel B, Metzger M, Blankestijn PJ, et al. Predicting kidney failure from longitudinal kidney function trajectory: A comparison of models. PLoS One. 2019;14(5).

219g Monte Carlo Simulations of Spherocylinders Interacting with Site-Dependent Square Well Potentials

Kiranmai Yellam, Prateek Jha

Monte carlo simulations are performed to study Self assembly of spherocylinders

Monte Carlo Simulations of Spherocylinders Interacting with Site-Dependent Square-Well Potentials

Yellam Kiranmai, Prateek K. Jha

Department of Chemical Engineering, IIT Roorkee, Roorkee, Uttarakhand – 247667, India

Monte Carlo Simulations are performed to study the self-assembly of spherocylinders with square-well interactions between some sites on the spherocylinders. Simulations are performed on a homogeneous system comprised of identical spherocylinders and a binary system comprised of spherocylinders of different lengths and interaction strengths/range. The spherocylinder concentrations, interaction strength and range of interactions are varied to develop the phase diagram of the systems. We observe the formation of dispersed, semibundled, bundled, and network configurations of the system depending on the choice of model parameters. The square well potential represents hydrogen-bonding interactions when the interaction strength and range is appropriately chosen. This model therefore mimics the self-assembly behavior of hydrogen-bonding rodlike molecules, e.g., short-length oligomers. The bond formation and breakage is further analyzed by the calculation of bond lifetimes from the time correlation function. Some of the results of the study can be compared with previous results of the atomistic simulations performed on short oligomers with/without drug molecules.

89a Low-Level Metal Toxicology in Serum Via Anodic Stripping Voltammetry

Shaylee R. Larson, Mary Jeppson, Kevin Chandler, Paige Leland, Himanshu Sant, Swomitra Mohanty

In recent years, the risk of exposure to heavy metals has risen globally due to increased anthropogenic processes and rapid industrialization.[1] To this effect, the World Health Organization (WHO) reports that deaths caused by exposure to hazardous chemicals rose 29% between 2016 and 2019.[2] Though preventable, toxic metal exposures in remote or medically-poor areas go largely underreported due to the lack of rapid and inexpensive testing[3]. Electrochemical methods of metal detection have often been used for their sensitivity in water, however, their capacity directly in complex biological samples has yet to be extensively explored. In this work, detection of copper (Cu) and lead (Pb) metal in serum was conducted via anodic stripping voltammetry (ASV). The developed rapid serum-digest method allows for results within 4 hours of sample collection time without losses in sensitivity. Stripping parameters such as deposition potential and scan rate were optimized to allow for detection at physiologically-relevant concentrations. Lower limits of detection in serum were experimentally determined to be within the physiologically relevant range (ppm). Given these outcomes, the sensing platform provides the basis for low-cost, rapid point-of-care toxic metal detection allowing frequent testing to become more accessible for vulnerable populations.

Citations

(1) Elsevier. Heavy Metals in the Environment: Origin, Interaction and Remediation,

1st ed.; Bradl, H., Ed.; Academic Press Inc. (London): London, England, 2005.

(2)New data on the public health impact of chemicals: knowns and unknowns

https://www.who.int/news/item/06-07-2021-new-data-on-the-public-health-impact-ofchemicals-

knowns-and-unknowns (accessed 2022 -04 -11).

(3)Rees, N.; Fuller, R. The Toxic Truth: Children's Exposure to Lead Pollution

Undermines a Generation of Future Potential the Toxic Truth: Children's Exposure to

Lead Pollution Undermines a Generation of Future Potential.

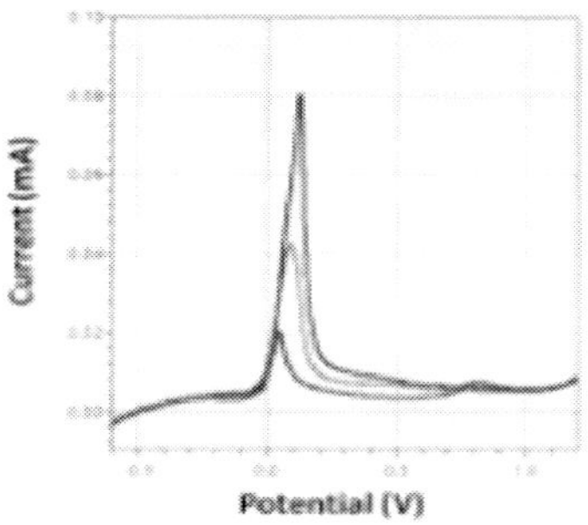

Anodic stripping peaks for Cu in serum. Serum sample was spiked at 10 ppm (purple), 5 ppm (green), 1ppm Cu (blue, 2700 s dep time), and digested prior to testing. Note after dilution from the digestion process, the concentration of Cu in the sample was 1ppm, 0.5ppm.

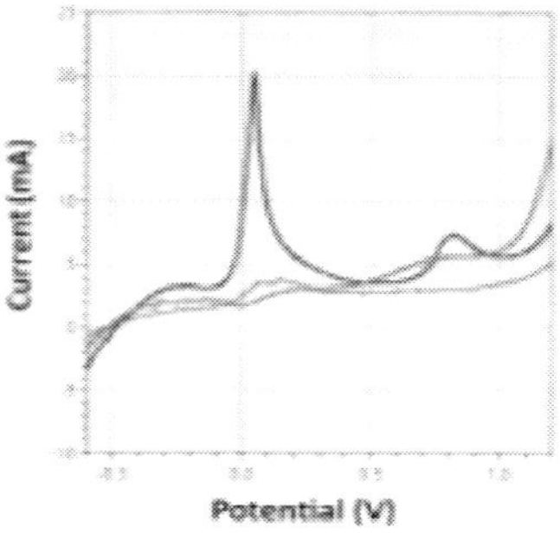

Anodic stripping peaks for Cu at 1 ppm (blue, 2700 s dep time), 0 ppm (green) Cu in serum, Acetate buffer (red). Results show that minimal signal is present at the peak voltage in negative control samples.

89b Comparison of Methemoglobin and Deoxyhemoglobin As Potential Contrast Agents in Magnetic Resonance Imaging (MRI)

Royaalsadat Ayati, Kyle Manwaring, Randy S. Lewis

Magnetic Resonance Imaging (MRI) is a valuable tool for the diagnosis of a patients' health condition. Contrast agents, which are paramagnetic, are used to enhance the signal intensity (brightness) of the image. Gadolinium-based contrast agents are broadly used due to their effective ability to enhance the MRI signal intensity. However, concerns about the limitations and safety of gadolinium-based contrast agents, such as accumulation in tissues, have led to an interest in alternative contrast agents.

Methemoglobin and deoxyhemoglobin are naturally occurring blood species that have the potential to be used as contrast agent due to their paramagnetic properties. Previous studies in humans (deoxyhemoglobin) and dogs (methemoglobin) have shown that these two species can effectively enhance MRI signal intensities. However, it is unknown as to which of these species has the better ability to be used as a contrast agent. Methemoglobin is typically 1-2% of the total hemoglobin. It has been shown that up to 20% of methemoglobin can be tolerated. Methemoglobin can be produced via the addition of nitric oxide (NO) or nitrite.

This work involved the development and analysis of a NO-delivery system to rapidly produce methemoglobin. The system could also produce deoxyhemoglobin. Methemoglobin was produced by exposing oxyhemoglobin to NO using gas permeable tubes. Deoxyhemoglobin was produced by eliminating oxygen in the head space of a blood reservoir to force oxygen to be removed from the blood. Results showed that methemoglobin is essentially the only product produced during NO delivery until near complete conversion of oxyhemoglobin to methemoglobin. Methemoglobin is stable if NO delivery is removed at this point. However, if NO delivery is continued after making methemoglobin, nitrosylhemoglobin species are also produced.

As methemoglobin or deoxyhemoglobin were made, samples were analyzed with an MRI. The T1-weighted measurements were obtained for these samples. In some cases, a mixture of methemoglobin and nitrosylhemoglobin were also analyzed for the TI-weighted measurements. The T1 values were compared among the various studies to identify which blood species had the best potential to be used as a contrast agent. Initial studies have shown that methemoglobin is the most viable hemoglobin-based candidate to be used as a contrast agent.

89c Barrier-Free Paper Analytical Devices for Multiplex Colorimetric Detection

Ayushi Chauhan, Bhushan Toley

Microfluidic paper-based analytical devices (μPADs) are being increasingly adopted to perform colorimetric assays for various applications. The reason is attributed to their portability, biocompatibility, and pump-free designs. Multiplex colorimetric detection in μPADs is achieved by patterning hydrophobic barriers (Martinez *et al.*, 2007). These barriers were created using wax printing, inkjet printing, photolithography, or chemical modification of paper. Although these fabrication methods efficiently utilize membrane surface area and consume less reagent volume, they require expensive equipment and skilled personnel, posing a challenge for scale-up. In addition, the convection of rehydrated reagents towards membrane edges produces non-uniform signals, which make result interpretation difficult. To overcome the limitations of conventional μPADs, we present a new device called barrier-free μPAD (BF-μPAD) that can detect multiple targets without patterning barriers, enabling faster and homogenous signal generation. In one embodiment, an 8cm x 2cm device assembly can perform up to 20 different tests in 30 seconds. A user only requires a smartphone to visualize and interpret signals, making it a desirable point of care tool at low resource settings.

BF-μPADs consists of a stack of two paper layers of different wicking rates. The top layer is composed of Standard 17 glass fiber (called distribution layer) and offers less resistance to fluid flow. The bottom layer is composed of Whatman filter paper Grade 1 (called detection layer) and offers high resistance to fluid flow. The assembly was secured using acrylic supporting layers. Test reagents were deposited on the detection layer, and samples containing different analytes were added through the distribution layer. After the signals were fully developed, the device was flipped, and the colored signals were visualized using a smartphone in a custom-made imaging box and using a flatbed scanner. The image was then split into RGB color channels, and intensities of each spot were measured in the suitable channel using ImageJ software.

Prior to multiplex detection, a preliminary study was performed to investigate the uniformity of signals on the detection layer. 16 spots of bromophenol blue, a pH indicator dye, were deposited on an 8cmX2cm detection layer (Fig. 1A). An equal dimension of Standard 17 was placed flush on the detection layer, and 895ul (volumetric capacity of distribution and detection layer) of 1M HCl was added as the rehydrating fluid through the distribution layer. On addition of the acid, the pH of the dye changes, and the blue spots turn yellow. Figure 1B shows the colorimetric signals developed in the detection layer. It was observed that the intensities of the yellow color were similar and independent of the location of spots from the fluid source (Fig. 1C). The coefficient of variations of 48 spots across three device replicates was 6.94% for the smartphone and 3.49% for the scanner.

The multiplexing feature of BF-μPAD was demonstrated by testing salivary thiocyanate, glucose, nitrite, and bovine serum albumin (BSA) in a 2cm x 2cm assembly. Figure 2A shows the user experience of the device. After saturation of signals, the device was flipped, and the image of the detection layer was acquired using a smartphone. Figure 2B shows the colorimetric signals obtained in each spot for a sample containing 1 mM SCN^-, 10 mg/dl glucose, 0.8 mM NO_2^- and 3 mg/ml BSA. Eight sample fluids with different concentrations of the four analytes were added to BF-μPAD, and mean test spot intensities (N=3 devices) were plotted against analyte concentrations. Intensities of all analytes increased monotonously with concentrations. Figure 2C shows the calibration plots of the four analytes.

We also compared the limit of detection (LOD) of the BSA assay performed in BF-μPAD and a traditional single-layer μPAD using a statistical method described by Holstein et al. (Holstein *et al.*, 2015). LOD and the corresponding 95% confidence interval obtained using BF-μPAD were 0.1817 mg/ml and 0.1283-0.3682 mg/ml, respectively. The corresponding figures for μPAD were 0.6727 mg/ml and 0.5314-0.8217 mg/ml, respectively, signifying that BF-μPAD improved the LOD of the colorimetric assays by 3.7x compared to conventional μPAD. Further, a comprehensive study was performed to investigate the cross mixing of rehydrated signals on a longer assembly by depositing additional reagents upstream of the test signals. It was observed that the rehydrated signals from one chemistry do not mix with the downstream signals.

Barrier-free colorimetric detection in BF-μPAD is achieved by the unique flow pattern. The distribution layer rapidly wicks fluid in the lateral direction and continuously supplies it to the bottom detection layer from the vertical direction. Vertical entry of fluid practically eliminates lateral fluid movement in the detection layer, obviating the need for patterning barriers and generating spatially uniform signals. These flow patterns were modeled in COMSOL Multiphysics using Richards equation but are not presented here for brevity. Although the application of BF-μPAD was shown for biological fluid, the device can also be adapted to detect different pollutants in the environment, contaminants in soil, and pesticides in food. Since the distribution layer has large pores, the assembly is well suited for testing viscous emulsions like milk and denser fluids like honey which are challenging to flow alone through the cellulosic membrane of conventional μPADs.

BF-μPADs avert the need to pattern paper and enable faster flow and generate spatially uniform signals, improving upon the LOD of the assay. In contrast to μPADs, large-scale manufacturing of BF-μPADs would only require a commercially available benchtop robotic dispensing system to handle microliter volume.

89d Clinical Validation of Breath Biomarkers in Pediatric Tuberculosis Patients Using Comprehensive Two-Dimensional Gas Chromatography Time-of-Flight Spectrometry

Mary Jeppson, Emily Lym, Swomitra Mohanty

Background and Motivation

Current methods used for Tuberculosis (TB) testing are expensive and take days to weeks to provide diagnosis. Tuberculosis is widespread in the developing world and is the second leading cause of hospitalizations in the United States. The Advanced Materials and Microdevices Lab at the University of Utah is currently working on sensor applications aimed at designing a low-cost point-of-care sensor to improve access to tuberculosis testing in the developing world. These sensors are specifically designed to target tuberculosis volatile organic compounds (VOCs) in breath.

Many human diseases have corresponding breath VOC profiles and breath analysis is increasingly the focus of new diagnostic techniques. This diagnostic approach is non-invasive and can serve as a safer, less time intensive diagnostic tool than blood or sputum tests. A number of diseases have been found to have breath profiles unique to the disease; such as influenza, diabetes and pneumonia.

While two VOCs, namely methyl-nicotinate (MN) and methyl-p-anisate (MPA), have been uniquely identified as VOC indicators of active tuberculosis disease in adults, the data to date is inconclusive in the pediatric population.

With the aim of identifying more robust diagnostic biomarkers in children, the University of Utah in partnership with Makerere University in Uganda and the University of California, San Francisco are conducting a large VOC breath study with over 700 patients enrolled as of April 2022. A number of biomarkers correlating to active TB are being identified using two-dimensional gas chromatography, Time-of-Flight spectrometry (GCxGC TOF).

Methods

Breath samples are extremely complex. While many industries employing GC-MS are concerned with a handful of constituent sample elements, disease diagnostics must consider thousands of chemically similar trace-level endogenous and exogenous metabolic byproducts. While a well-designed sample acquisition method can employ chemical and thermodynamic principles to achieve reductions in inevitable peak convolutions, two-dimensional chromatography is capable of separating compounds using contrasting column stationary phases. This enables the magnification of small chemical differences and provides access to more trace compounds than could be seen with traditional GC-MS. Additionally, the TOF mass spectrometer provides a full-spectrum scan as opposed to a quadrupole resulting in no loss of ion data.

With greater volumes of data acquired using these technologies and methods, comes the need for improved data processing to make sense of it. We have developed data processing methods using the Chromcompare+ software from Markes International.

Finally, after likely biomarkers are identified, the compound identification must be validated by the use of known standards. The behavior of these known compounds must be compared to the ion signature of the suspected TB biomarker. This requires recognizing subtle differences in mass spectral profiles.

Significance

Tuberculosis kills millions of people unnecessarily every year. With the development of an inexpensive, real-time breath sensor, this would not have to continue. The foundational step in the sensor technology development is identifying a full breath profile of both pediatric and adult TB patients. This full breath profile will be used to create the first diagnostic breath sensing device with a sufficient sensitivity and specificity to achieve FDA approval.

This presentation will discuss each of these new developments in the application of GCxGC TOF systems to the breath biomarker validation process including the application of chemical engineering principles to complex sample separation and identification, how we arrived at optimized data processing using Chromcompare+ and newly discovered TB biomarkers.

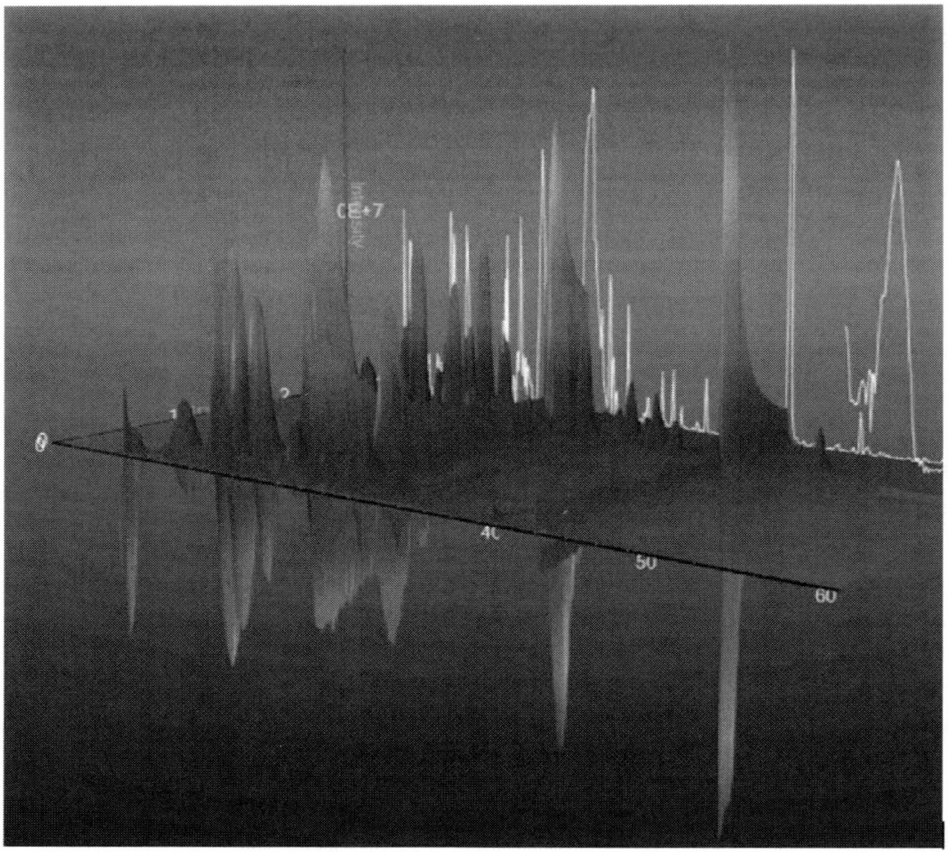

Figure 1: Two-Dimensional total ion chromatogram. Two patient samples are superimposed and one is inverted to identify differences visually.

89e Elucidating the Effects of Shear and Surface Topography on Thromboembolism in Ventricular Assist Devices

Anjana Jayaraman, Junhyuk Kang, James Antaki, Brian J. Kirby

Continuous-flow left ventricular assist devices (VADs) are implanted in end-stage heart failure patients to assist with blood circulation. However, VAD implantation often causes thrombosis or blood clot formation, which results in complications such as device malfunction, invasive VAD replacement surgeries, stroke, and death. Several studies over the past three decades have elucidated components of VAD implantation that contribute toward thrombus formation, which have been famously defined as Virchow's triad, that is, blood-surface contact, blood flow dynamics and innate biochemical factors impacting blood coagulability.

When blood contacts implanted surfaces, which in the case of VADs are often made of titanium alloys or polymers, plasma proteins adsorb to the surface, creating an adhesive layer for platelets i.e., thrombocytes to bind to with surface receptors.Therefore, implanted surfaces serve as a nidus for the formation of potentially dangerous thrombi, comprised mainly of platelets and a network of plasma proteins such as fibrinogen and von Willebrand factor (vWf).

Thrombosis is driven by the aggregation of active platelets with plasma proteins and each other. Platelets may exist in either resting or activated states, the latter of which have higher protein binding affinities due to increased expression of surface receptors. With these receptors, platelets bind to key free-flowing proteins, fibrinogen and von Willebrand factor (vWf), which exist in globular conformations when freely flowing in blood. Shear rates as low as 50 s^{-1} induce fibrinogen unfolding and expose binding sites for platelets. However, due to the weak nature of the fibrinogen-platelet receptor bonds, the number of platelets binding to fibrinogen markedly drops after 1000 s^{-1}. On the other hand, vWf is a larger multimeric protein that elongates only at shear rates above 1000 s^{-1}, with the largest multimers elongating at 5000 s^{-1}. vWf forms stronger bonds with platelets than fibrinogen, therefore regions above 5000 s^{-1} support stable platelet aggregates. These well-established findings highlight the vital biophysical role played by physiological shear rates (<2000 s^{-1}) and non-physiological shear rates (>2000 s^{-1}) that are often applied by VAD pumps.

Some research has been done to elucidate the impact of varying surface topographies on platelet adhesion and/or surface protein adsorption; however, the conclusions are widely distributed and contradictory. Reported results range from a 150% decrease to a 400% increase in platelet adhesion on sub-micron scale roughened surfaces. This could be attributed to many parameters in the studies, ranging from roughness characterization to the scales or aspect ratios of topographical features. A majority of studies we found report roughness in terms of either arithmetic mean height (Ra) or contact angle. While the former indicates feature size in a simplified manner, the latter can be misleading on surfaces with nanoscale features. Furthermore, some studies were performed under static conditions whereas others involved hemodynamic flow conditions, thus not capturing flow-induced protein interactions with both the surface and platelets.

To that end, this work was designed to determine the impact of micro- and nanoscale titanium alloy surface topographies on platelet adhesion and platelet aggregate properties under various physiological and VAD-relevant shear rates using microfluidic devices. We used fluorescent microscopy to visualize surface adhesive platelets in real time, which were then quantified using an image processing algorithm. Additionally, part of the study was aimed toward systematically describing surface roughness using surface spatial frequencies to make the findings adaptable to more blood- biomaterial contact applications.

Methods

To investigate the phenomenon of shear-induced thrombosis on various surface topographies, we incorporated a set-up with a polydimethylsiloxane (PDMS) microfluidic device clamped on to a roughened titanium-alloy surface. Polished Ti-alloy surfaces were roughened to arithmetic mean height (Ra) values of 0.3, 0.8 and 2.1 μm using abrasive and sandblasting techniques. A mirror-finished surface with no additional polishing was used as a negative control. Three-dimensional surface profiles were obtained via laser scanning profilometry, and a custom MATLAB program was written to calculate the Ra values and generate mean Fourier transform energy profiles. We designed and fabricated microfluidic devices to precisely control the flow conditions and surface chemistry, while visualizing thrombus formation at the micro-scale. The microfluidic channels had rectangular geometries with low Reynolds and high Peclet numbers so as to apply constant, well-defined wall shear rates over the length of the channel. Fluid shear rates of 1000, 2000 and 5500 s^{-1} were applied by varying the input flowrate, controlled by a syringe pump.

The perfusate was a blood analog comprised of platelet-rich plasma and hemoglobin-depleted red blood cells, thereby creating an optically clear suspension with similar composition and rheological properties as whole blood. The platelets were labeled with a fluorescent dye named mepacrine (quinacrine dihydrochloride), which combined with the suspension's optical clarity, enabled the visualization of platelet adhesion and aggregation, and thus thrombus formation, via upright fluorescent microscopy over 15-minute durations. The entire system was maintained at a temperature of 37°C to mimic physiological cellular mechanotransducive responses.

Images acquired via fluorescent microscopy were computed using a custom MATLAB program that included flat-field illumination correction, noise removal using pixel-wise low-pass temporal filtering, and binarization using Otsu dual-thresholding methods. From the binarized images, object (aggregate) properties were obtained, namely individual aggregate area, number of aggregates per frame and eccentricity of aggregates (not shown here). Additionally, mean object fluorescence intensities were plotted to compare aggregate densities across the parameter space (fig 2).

Results

At 1000 s^{-1}, adhesion patterns demonstrated sustained growth of platelet monolayers (fig 1a) when Ra<2.1 μm. At 2000 s^{-1}, stable, growing platelet aggregates were observed when Ra<2.1 um (fig 1b). At 5500 s^{-1}, round platelet aggregates formed on all three surfaces, often connected by fibrous structures (fig 1c), with increasing intensity and density with time. Embolization was observed on all surfaces after an initial increase in aggregate intensity (in yellow, fig 2c). On the Ra 2.1 μm surface, platelet aggregates formed at all three shear rates, indicative of higher shear rates and gradients as a result of the roughness.

Future work will incorporate further analysis of platelet aggregate properties such as individual aggregate count, mean aggregate area and eccentricity. The former two properties determine the distinction between nucleation and growth of aggregates and the nature of aggregate growth; i.e., shape and direction of platelet deposition. Furthermore, these properties will be examined in conjunction with the spatial frequencies associated with the various surface topographies.

Implications

At both physiological and non-physiological shear rates, a surface with a 2.1 μm Ra causes thrombosis. Under non-physiological VAD-relevant shear, roughness leads to the embolization of thrombi, a phenomenon that has previously not been observed in studies investigating surface effects on thrombosis. These results demonstrate that rough surface topographies synergize VAD shear-induced thrombosis and thromboembolism. More broadly, these findings highlight the importance of polishing device surfaces to prevent dangerous and potentially fatal health complications.

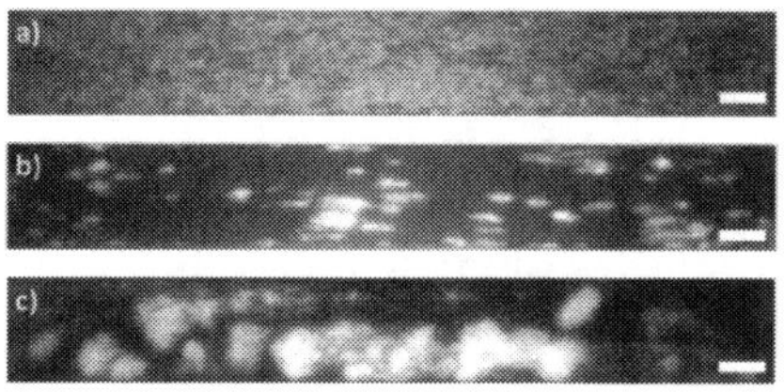

Figure 1. Scale bar length is 100 μm.

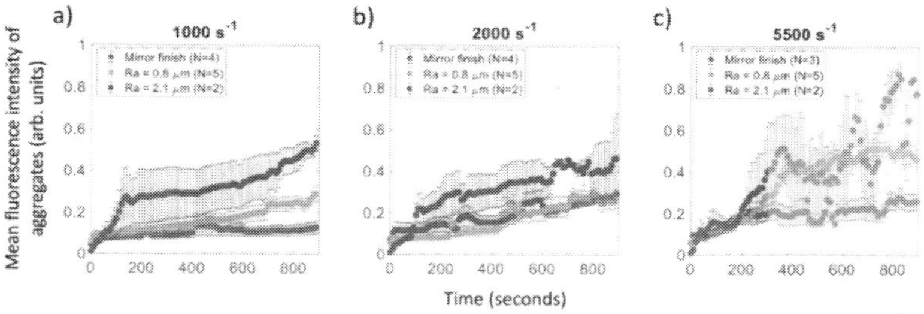

Figure 2. Mean fluorescence intensity of aggregates. Error bars represent standard error of the mean at each timepoint.

89f Active Levy Swimmers and Geometric Design for Anti-Infection Catheters

Tingtao Zhou

Antibiotic resistant bacteria pose a serious threat to hospital patients who are already immunocompromised. Recent experiments have demonstrated upstream swimming and super-contamination caused by bacteria inside microfluidic channels, which has been suspected to be responsible for many urinary infection cases. In the first part, our theoretical work on Active Levy Swimmers (ALS) addresses the dynamics of very persistent bacteria within channels. In contrast to the classical model of Active Brownian Particles (ABP), ALS shows fundamentally different non-equilibrium properties that we prove to follow a fractional-order partial differential equation (FPDE). For example, Gaussian diffusion leads to uniform steady state distributions in the bulk, while our FPDE and particle dynamics simulations show a power-law bulk distribution for ALS. Based on the theoretical understanding of the super-contamination mechanism, we propose a novel geometric design for anti-infection catheters inspired by the natural structure of bamboos. Our life-size experiments show that this design successfully disrupts the persistent upstream swimming of bacteria and significantly reduces upstream contamination. Our simulations and microfluidic experiments corroborate the life-size experiments and reveal the underlying mechanism of this design.

89g A Low Cost, Compact Flow Cytometer

Mahrukh Mir, Mahesh S. Tirumkudulu

Flow cytometry is a powerful tool for measuring microscopic or sub-microscopic particles or biological cells. The technique enables measurement of physical, chemical and biological properties of cells, micro-organisms, and particles. Laser light is scattered from particles focused in a flow cell and collected by light sensors (such as photodiodes), where the intensity of the scattered light is a function of the scattering angle, refractive index of the particle and surrounding medium, wavelength of light, and the size and the shape of particle.

One of the critical parts of the cytometer is the flow cell where the particle stream is constrained into a tight region using hydrodynamic focusing. The conventional flow cells use thick quartz flow cells, they are expensive and therefore not suitable for instruments targeted for resource-constrained settings. We propose a low-cost sheath flow assembly design made of micro-tight ETFE adapters and an assembly of capillary tubes for the passage of focussed flow of particles. The entire flow cell assembly is a simplified version of a sheath flow cuvette design that is able to focus and accelerate a stream of particles in a narrow region probed by the incident laser light. The flow cell is embedded in a 3D-printed module, which houses microlenses and photodetectors, resulting in a compact flow cytometer. We show excellent agreement between the size distribution of model particles obtained via direct imaging and those obtained from light scattering. We have been able to obtain a 3-part differential of WBC based on the forward scatter measurement on our device. With minor modifications, the device may be used for disease detection, such as targeting viruses and bacteria.

The attached figure compares the size distribution of two different sets of particles obtained from image analysis of microscope images and that obtained from measurement of forward scattering data. The figure shows, (a) an overlap between the normalized Gaussian distribution of the cross-sectional area of 14-20m particles, (⁄)^2, and the normalized Gaussian forward scattered intensity distribution,⁄. Here, is the mean value of the diameter of the particles in the mixture while is the mean intensity value in the scattered light intensity distribution. The forward scatter intensity follows a direct square relationship with the particle diameter. A microscope image of the 14-20m particles in suspension has been provided in the inset for reference, (b): An overlap between the normalized Gaussian distribution of the cross-sectional area of 8-12m particles, (⁄)^2 and the normalized Gaussian distribution of the forward scattered intensity,⁄. A microscope image of the 8-12m particles in suspension has been provided in the inset for reference.

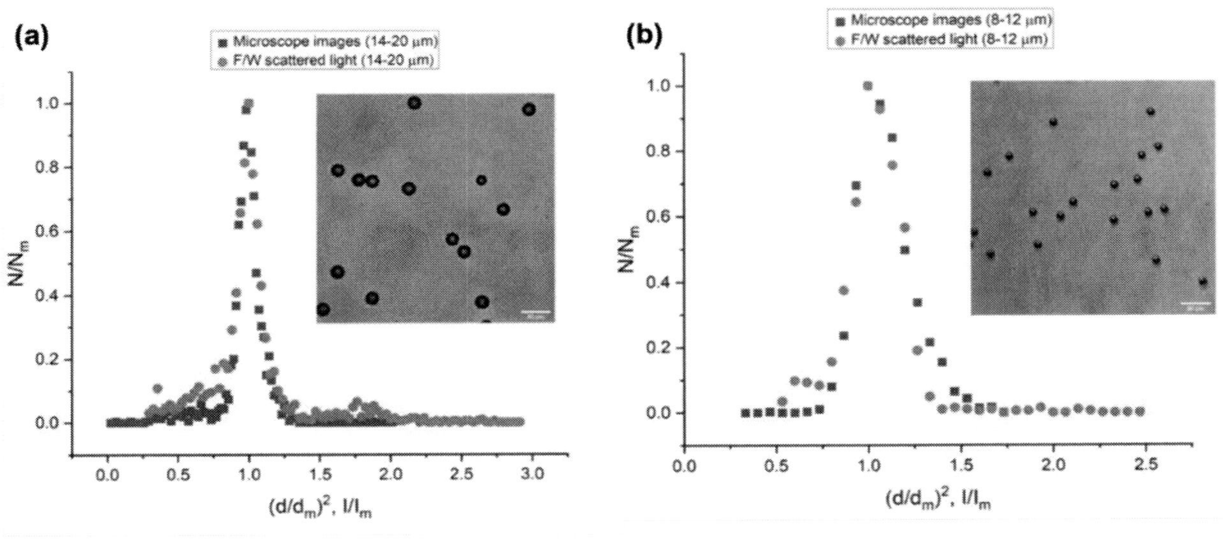

153a Slow-Release Nanoparticles for Chronic Pain Management

Parker Lewis, Rachel Pollard, Rocco Latorre, Dane Jensen, Brian Schmidt, Nigel Bunnett, Nathalie M. Pinkerton

A reliable, non-habit-forming treatment for chronic pain has been elusive due to the rapid bodily clearance and short therapeutic effects of traditional opioids. To reduce opioid dependence, there is a need for a therapy which achieves robust pain inhibition for extended periods of time. Recent studies have shown that delivering drugs directly to spinal neuron endosomes yields more potent inhibition of key pain-signaling pathways. Additionally, nanomedicine has shown promise in promoting endosomal uptake for anticancer therapeutics. To translate this strategy to pain therapy, it must be demonstrated that antinociceptive drugs can be efficiently encapsulated into nanoparticles, while controlling their release rate to prolong drug activity. Here, we develop a polymeric nanoparticle drug delivery platform for the sustained release of pain-signaling antagonists in spinal neuron endosomes.

Hydrophobic ion pairing (HIP), a method of solubility tuning, has been shown to promote the encapsulation of small molecule drugs within polymeric nanoparticles. We hypothesized that employing hydrophobic counterions during nanoparticle assembly would both increase the encapsulation efficiency of a calcitonin gene-like receptor (CLR) antagonist and slow the diffusion rate of this antinociceptive drug from the nanoparticle core.

The CLR antagonist and hydrophobic counterions were encapsulated by amphiphilic block copolymers via Flash Nanoprecipitation (FNP), a rapid and size-tunable nanoparticle assembly process. Synthesis parameters including counterion species, drug/counterion molar ratios, and solvent systems were varied to optimize nanoparticle drug loading (wt.%) and *in vitro* drug release rates.

Pairing of the CLR antagonist with pamoic acid resulted in nanoparticles (100 nm in diameter) with drug loadings of up to 6 wt.% and encapsulation efficiencies of up to 80%, with stability observed for over two weeks. These formulations demonstrated sustained drug release in sink condition release media for over nine days.

The reproducible encapsulation of an antinociceptive drug demonstrates the ability for endosomal uptake by spinal neurons, preventing rapid clearance. In achieving high drug loading compositions and sustained drug release on the order of weeks, such formulations are optimal for the potent and prolonged inhibition of the CLR pain-signaling pathway, presenting a reproducible framework for chronic pain therapies.

153b *Pseudomonas Aeruginosa* Reverse Diauxie Is a Multidimensional, Optimized, Resource Utilization Strategy Facilitating Chronic Wound Colonization

S. Lee McGill, Yeni Yung, Kristopher A. Hunt, Michael Henson, Luke Hanley, Ross P. Carlson

Pseudomonas aeruginosa is a globally-distributed bacterium often found in medical infections. The opportunistic pathogen uses a different, carbon catabolite repression (CCR) strategy than many, model microorganisms. It does not utilize a classic diauxie phenotype, nor does it follow common systems biology assumptions including preferential consumption of glucose with an 'overflow' metabolism. Despite these contradictions, *P. aeruginosa* is competitive in many, disparate environments underscoring knowledge gaps in microbial ecology and systems biology. Physiological, omics, and *in silico* analyses were used to quantify the *P. aeruginosa* CCR strategy known as 'reverse diauxie'. An ecological basis of reverse diauxie was identified using a genome-scale, metabolic model interrogated with *in vitro* omics data. Reverse diauxie preference for lower energy, nonfermentable carbon sources, such as acetate or succinate over glucose, was predicted using a multidimensional strategy which minimized resource investment into central metabolism while completely oxidizing substrates. Application of a common, *in silico* optimization criterion, which maximizes growth rate, did not predict the reverse diauxie phenotypes. This study quantifies *P. aeruginosa* metabolic strategies foundational to its wide distribution and virulence including its potentially, mutualistic interactions with microorganisms found commonly in the environment and in medical infections.

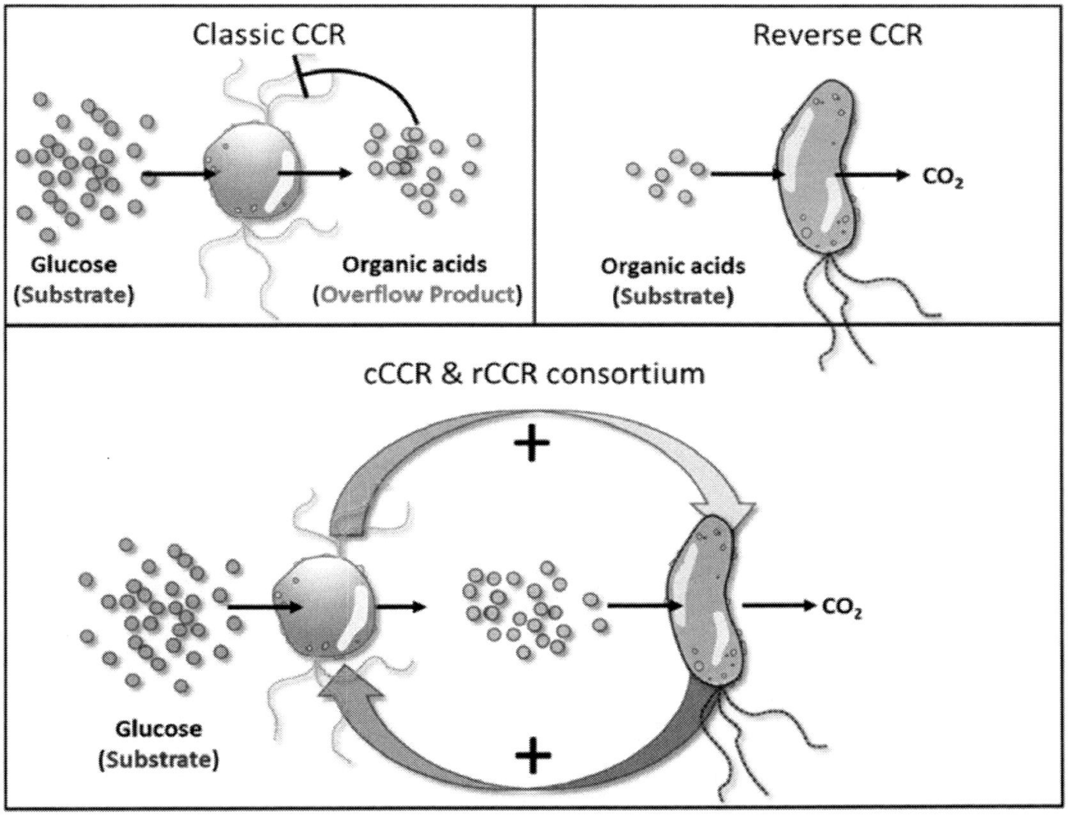

153d Evaluating Antimicrobial Drug Formulations on a 3D Biofilm Growth Model Under Simulated *In Vivo* Conditions

Yadiel Varela-Soler, Himani Chavda, Swarnima Roychowdhury, Veronica Farag, Patrick Erickson, Biju Parekkadan, Charles Roth

Bacterial biofilms are defined as aggregates of bacteria that are embedded within a self-produced matrix, which consists of extracellular proteins (e.g. actin), DNA, and exopolysaccharides (most notably alginate). It has been demonstrated that formation of these significantly increase the risk of persistent chronic infections due to increased antibiotic resistance, specifically in the setting of cystic fibrosis (CF), where thick mucus acts as a secondary matrix. Traditional microbiology assays consist of 2D biofilms in well plates under static conditions or mild agitation and thus do not represent well the natural environment of biofilm infections. In vivo infection models often utilize beads made from alginate, a natural component of bacterial biofilms, as a substrate to initiate biofilm formation. Alginate beads could provide a more biomimetic 3D biofilm growth model for in vitro studies as a bridge to animal models.

In this study, biofilms formed in alginate beads were incorporated into a novel engineered system where the biofilms can be exposed to pharmacokinetically relevant drug profiles. The system comprises of an engineered perfusion bioreactor where the alginate beads containing *P. aeruginosa* (PA) are incubated and exposed to different drug profiles. Results demonstrate that viable PA, at concentrations well above $1x10^7$ CFU/bead, were successfully embedded within alginate beads as determined by colony forming unit (CFU) assays. Treatment of the beads with the gold standard antibiotic, tobramycin, at clinically relevant concentrations and an exposure time of 2 hours showed no effect in terms of cell growth, detachment of cells from biofilm, or glucose uptake. However, longer exposure times of over 9 hours, at the same concentrations, produced a significant reduction in glucose consumption and bacteria release from biofilm. These results highlight the importance of drug concentration profile on killing of alginate-embedded biofilms and suggest that our proposed system can serve as a bridge between in vitro and in vivo experiment. Current efforts are being made to test additional exposure times, bacteria strains, antimicrobials, and pharmacokinetic profiles.

153f Improving Therapeutic Protein Secretion in Probiotic Yeast *Saccharomyces Boulardii* using a Multifactorial Engineering Approach.

Deniz Durmusoglu, Ibrahim Al'Abri, José Luis Martinez Ruiz, Nathan Crook

Recent advancements in synthetic biology enabled emergence of engineered live biotherapeutic products (LBPs). Engineered probiotics, as a class of engineered LBPs, are designer microbial therapies to target gastrointestinal disorders by delivering therapeutic molecule at the disease site [1]. Yeast as a eukaryotic cell factory poses several biomanufacturing advantages over bacteria for in situ drug delivery applications, such as ability to secrete high titers of protein and post-translational modification machinery. Particularly, probiotic yeast *Saccharomyces boulardii* (Sb) is a promising engineering chassis for next-generation probiotics applications such as in situ biomanufacturing of small molecules and therapeutic proteins in the gut [2,3].

In this work, we investigated means to improve recombinant protein secretion in *Sb*. We used 2 anti-*Clostridioides difficile* peptides and 4 anti-*Clostridioides difficile* nanobodies as model proteins. To validate and quantify secreted therapeutic peptide in the supernatant, we used SDS-PAGE and ELISA. After validating the secretion of recombinant proteins, we harnessed 3 strategies to tune protein secretion levels in engineered *Sb*. First, we investigated the effect of expression vector on the protein concentration in the supernatant. We compared genome-integrated vectors and plasmidic vectors with low- and high-copy origins. Second, we investigated the effect of secretion signals on therapeutic protein concentration. We screened 6 signal sequences, 3 native signals and 3 engineered signals. Thirdly, we engineered the major steps in the secretory pathway in *Sb* to improve the therapeutic protein secretion. We created a knock-out library of 17 Sb strains via CRISPR-Cas9 genome editing and screened for protein production and growth using microbioreactors. Lastly, we investigated the *in vivo* colonization and excretion profiles of engineered *Sb* strains in mice models. Through this engineering effort, we identified *Sb* strains with over 10-fold higher protein secretion levels than wild-type *Sb*, and over 4 orders of magnitude higher secretion levels than probiotic bacteria strains. Overall, this work lays the foundation for tuning recombinant protein secretion in *Sb* and establishing *Sb* as in-gut therapeutic biosynthesis platform.

153g A Low-Cost Strategy for Genotypic Antimicrobial Resistance Detection Using Oligonucleotide Ligation Assay

Ayushi Chauhan, Bhushan Toley

Infectious pathogens are known to accumulate point mutations (single nucleotide polymorphisms; SNPs) that enable them to spread more efficiently, evade the immune response, and develop resistance to drugs (antimicrobial resistance; AMR). The gold standard to determine point mutations is culture-based phenotypic antimicrobial susceptibility testing (AST). In this method, the presence or absence of resistant strain is determined by the growth or inhibition of the pathogen cells in a culture media containing the specific drug. While the method is accurate and inexpensive, it has a long turnaround time. The other method for AMR detection is the genotypic AST performed using nucleic acid amplification tests (NAATs). Identifying the genotype of the infection-causing agent or SNP-typing is essential in determining the proper course of clinical action, e.g., appropriate drugs, strategies for quarantining patients, etc. However, the current methods of SNP-typing are primarily restricted to whole-genome sequencing, which is infrastructure-intensive and cannot rapidly be scaled up. To meet the goal of a rapid and accurate genotypic detection method, we have developed an affordable and potentially scalable SNP-typing method that can be adopted at peripheral locations where the infection is most prevalent.

In this work, the point mutations are detected using oligonucleotide assay (OLA). OLA uses three differently labeled oligonucleotides or probes, two of which are allele-specific (Fig. 1A). When the probes hybridize to the amplified DNA, a ligase enzyme stitches the two ends of the probes and establishes a covalent bond (Fig. 1A). The formation of this bond is contingent on the presence or absence of mutation in the target DNA. OLA is a well-established method for SNP-typing but is currently restricted to certified laboratories. It has been used to detect point mutations in HIV using a well-plate-based immunochromatographic assay (Beck *et al.*, 2008, 2014; Panpradist *et al.*, 2016), in the *rpoB* gene of *MTB* using "DNA chips" (Jiao-Yu Deng *et al.*, 2004; J.-Y. Deng *et al.*, 2004) and in different genes of human and salmon sperm DNA using fluorescence resonance energy transfer (FRET) (Chen, Livak and Kwok, 1998). However, all these methods require expensive equipment and dedicated infrastructure. We have obviated these requirements by using a low-cost paper-based readout. As the ligated products of OLA are bi-labeled, it can be read on a lateral flow assay (LFA) akin to a home pregnancy strip. LFA consists of two test lines to detect the wild and mutant targets. For homozygous, a single test line would be seen either at WT or MUT, depending on the sample genotype. For heterozygous, relative intensities of the two test lines would indicate the ratio of wild and mutant strain in the sample. Prior to LFA, the ligation mixture is denatured to separate the unligated probes to prevent the generation of false-positive signals on the test lines. Further, dried premixes for PCR and OLA would be developed to reduce pipetting steps and user interventions. The method of deglycerolisation of enzymes radically improves the shelf life of dry-stored enzymes.

A potential application of the assay is in the rapid detection of drug resistance in the tuberculosis-causing bacteria, *Mycobacterium tuberculosis* (Mtb). Failure to identify specific mutations is the major cause of the spread of resistant strain and the high mortality rate of drug-resistant Tuberculosis (DR-TB) patients. WHO estimated that in 2018 there were half a million new DR-TB cases, majority of which were multi-drug resistant, i.e., resistant to at least rifampicin and isoniazid (first-line drugs for TB diagnosis). Out of these cases, only one-third were reported and diagnosed, highlighting the lack of an appropriate tool for the detection and surveillance of DR-TB. Current genotypic methods to detect DR-TB are NAATs followed by surface hybridization assays (line probe assays (LPAs) and GeneXpert) and real-time PCR assays (using molecular beacon probes). While GeneXpert and LPAs are rapid, their high operational cost restricts their use in certified laboratories. Similarly, real-time PCR requires sophisticated fluorescence imaging instrumentation. OLA, on the contrary, can be performed using low-cost portable PCR machines (e.g., Palm PCRTM and FranklinTM) by minimally trained users, potentially allowing same day selection of appropriate antimicrobial therapy.

Around 96% of mutations causing rifampicin resistance are conferred in the 81 bp rifampicin resistance detection region (RRDR) of the *rpoB* gene of Mtb. Out of which, the most prevalent mutation is at codon 531: TCG531TTG (S531L). The wild and mutant strains of Mtb having the 531 mutations were differentiated using OLA and LFA in 3 hours. The assay could detect 1 % mutation in a pool of wild and mutant synthetic DNA (Fig. 1B).

While the assay is being developed for DR-TB detection, the methodological and technological reformations witnessed during this project can be adapted to determine drug resistance in any pathogen. The assay thus will be a critical tool for managing infectious diseases in the future.

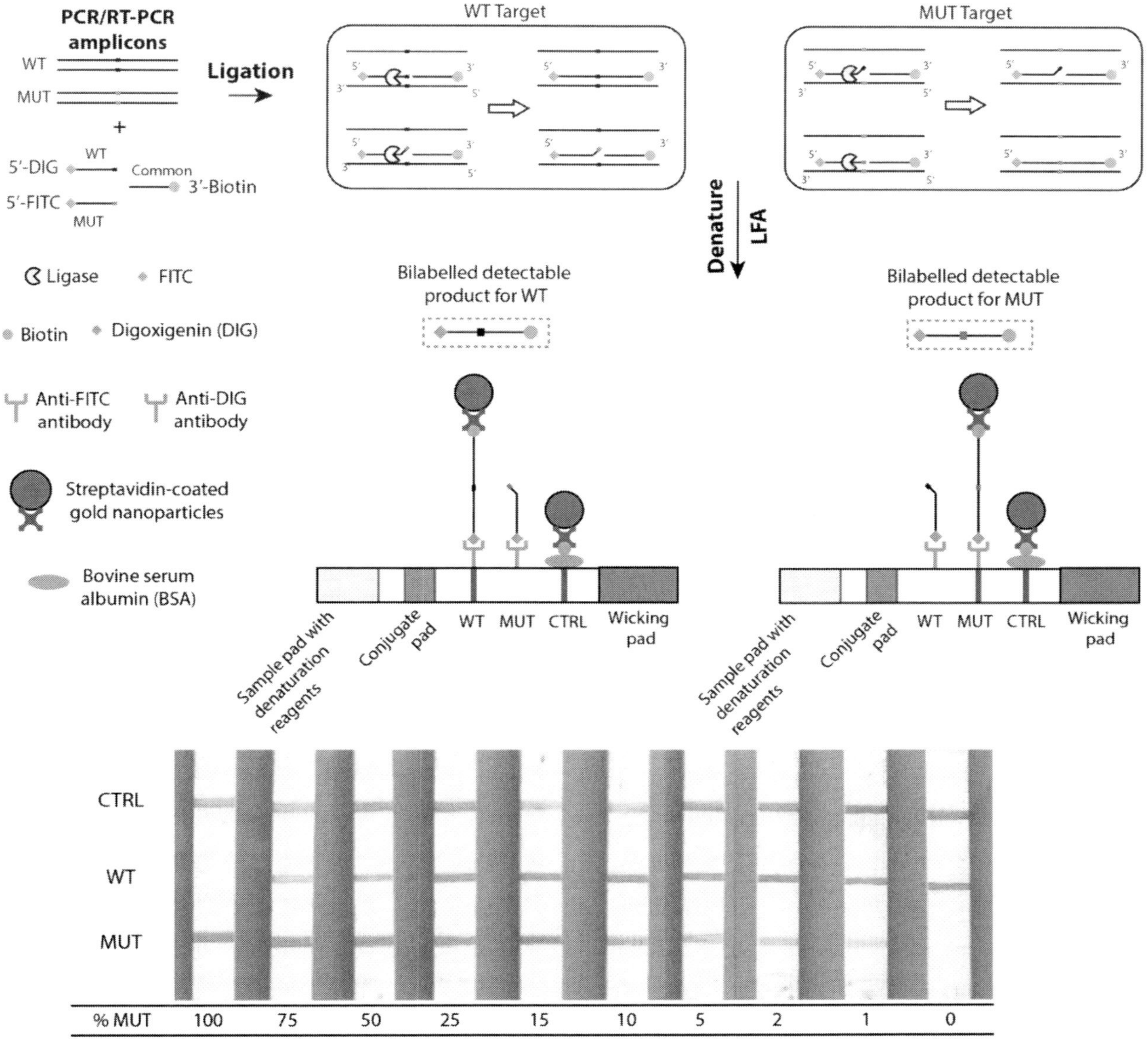

Figure 1. Detection of point mutation using OLA. (A) Molecular mechanism of the OLA assay. After denaturation, the bi-labelled single stranded products of WT and MUT target are detected on lateral flow assay. (B) Signal developed on WT and MUT test line with varying MUT % of the 531 codon of the rpoB gene of Mtb.

526a The Effects of Intratumoral Heterogeneity on Metastasis of Triple-Negative Breast Cancer Cells

Molly Brennan, Susan E. Leggett, Sophia Martinez, Celeste M. Nelson

Metastasis is responsible for most cancer-related deaths and evaluating the risk of metastasis is essential for predicting a patient's prognosis. Unfortunately, these predictions are challenging in the context of triple negative breast cancer (TNBC), a highly heterogeneous disease in which individual cells can undergo modifications at the genomic, epigenomic, transcriptomic, and/or proteomic levels. Over time, these modifications yield clonal subpopulations with distinct behaviors, giving rise to intratumoral heterogeneity. For instance, cells within a tumor may exhibit a spectrum of phenotypes due to the epithelial-mesenchymal transition (EMT). Our lab has derived clonal subpopulations from a mouse TNBC cell line with phenotypes that span the EMT spectrum. We have examined the behaviors of these clonal subpopulations individually and in mixtures at distinct steps along the metastatic cascade by taking advantage of engineered tumor models and an *in ovo* assay. Specifically, we have measured the ability of TNBC cells within engineered tumors to invade into the surrounding matrix and escape into a nearby cavity. Our data show that epithelial tumors invade and escape more slowly than mesenchymal or heterogeneous tumors do. Additionally, the presence of as few as one mesenchymal cell for every nine epithelial cells is sufficient to alter tumor morphology and increase the rate of invasion. The effects of mesenchymal cells are not due to paracrine signaling, since providing epithelial tumors with medium conditioned by mesenchymal tumors does not alter the rates of invasion or escape. To examine the metastatic capacity of the clonal subpopulations, we xenografted tumor cells onto the chick chorioallantoic membrane (CAM) and used a novel quantitative reverse transcription polymerase chain reaction-based method to detect murine cancer cells in the chick embryonic tissues. Using this engineered tumor model and *in ovo* model to examine the behaviors of homogeneous and heterogeneous cell populations has enhanced our understanding of the effects of intratumoral heterogeneity on metastatic potential. This understanding is necessary as physicians move towards more accurately predicting a patient's prognosis and developing personalized treatment plans.

526b The Effects of Palmitate-Induced IRE1α Activation in DNA Double Strand Break Repair and the Development of Chemotolerant Breast Cancer Cells

Kevin Chen, S. Patrick Walton, Christina Chan

Obesity is linked to increased risk of breast cancer, cancer progression, and chemtolerance. Although obesity is associated with reduced responses to chemotherapy, it is unclear how obesity worsens treatment outcomes and contributes to poorer survival rates in breast cancer patients. Given the increasing prevalence of obesity in the US population, understanding how obesity affects chemotherapeutic treatment is critical. Palmitate (PA), the most common saturated fatty acid in the human body, concentration is elevated in the blood of obese patients. Our laboratory previously showed that PA activates a pro-survival mechanism that promotes cancer progression and metastasis, mediated by an endoplasmic reticulum (ER) stress sensor protein - inositol-requiring enzyme 1α (IRE1α). There is evidence that supports the involvement of IRE1α in the DNA damage response. Previously our group showed that PA promotes the dimerization and activation of IRE1α. More recently, we found that PA reduced DNA double stranded breaks in breast cancer cells treated with a chemotherapy drug. Our results are the first to suggest a role of PA in DNA damage response mediated by IRE1α. Thus, we investigated the effects of PA on the DNA damage response of breast cancer MDA-MB-231 cells treated with a chemotherapeutic agent, etoposide (a potent inducer for DNA double stranded breaks). We hypothesized that the protective effect of PA on etoposide-treated MDA-MB-231 cells would depend on IRE1α. To test this hypothesis, we employed wild-type, IRE1α knockout, and IRE1α mutant MDA-MB-231 cells. Our results demonstrated that the activation of IRE1α by PA contributed to increasing DNA damage repair activity in response to etoposide-induced DNA double stranded breaks. This research is the first to uncover how PA contributes to DNA damage repair mediated through IRE1α, enhancing the survival of cancer cells treated with anti-cancer drugs. This is significant because (i) PA and ER stress are involved in numerous aging diseases, and (ii) a clear understanding of the molecular mechanism by which PA contributes to cancer cell survival will support the development of novel therapeutics for treating chemotolerance, which is more frequently detected in obese individuals.

526c Macrophages Aggregate and Cooperate to Phagocytose Cohesive Tumor Cell Targets in Engineered Immuno-Tumoroids and Solid Tumors

Lawrence J. Dooling, Jason C. Andrechak, Brandon H. Hayes, Siddhant Kadu, Dennis E. Discher

Macrophages are abundant in many solid tumors and are potential effector cells for monoclonal antibody-based cancer immunotherapies. This effect is dependent on Fc-mediated phagocytosis of cancer cells, but how macrophages engulf target cells that adhere to one another as in solid tumors is unknown. To address this, we engineered 'immuno-tumoroids' comprising soft but cohesive mixtures of melanoma cells with macrophages. Macrophages phagocytosed the cohesive melanoma cells only when an antitumor monoclonal antibody was supplied, and phagocytosis was maximized when signaling from the macrophage immune checkpoint ligand CD47 was abrogated by either CRISPR/Cas9 knockout or antibody blockade. Under the conditions for maximum phagocytosis, macrophages showed a striking spatial segregation within tumoroids. Although such segregation might be expected to hinder phagocytosis by limiting interactions between macrophages and target cells, we instead observed increased target engulfment by aggregated macrophages leading to tumoroid elimination. The results fit a phenomenological model of proliferation opposed by cooperative phagocytosis (i.e., Hill-like). Importantly, the same approaches for eliminating tumoroids in vitro also durably cured melanoma in immunocompetent mice when macrophage numbers were maximized by adoptive transfer of marrow-derived donor cells or engineered macrophages. Macrophage aggregates have previously been associated with increased survival in some human solid tumors, and our work provides a basis for engineering phase separating macrophages for maximum phagocytosis of cancer cells.

526d Mechanistic Characterization of Homologous Recombination in BRCA2 Mutant Cancers

Shayne Sensenbach, Prashant Karki, Mehmet Orman

The American Cancer Society estimates that over 40,000 women will die from either breast cancer or ovarian cancer in the United States in 2022 [1]. Mutations in the tumor suppressor genes, BRCA1 and BRCA2, lead to lifetime risks of 50-80% and 30-50% for developing breast and ovarian cancer, respectively [2]. BRCA mutations impair homologous recombination (HR), an important mechanism for repairing double-strand breaks (DSBs) in DNA. Cancer cells with BRCA mutations have deficient levels of the BRCA2 protein, but HR is still achieved through a RAD52-dependent backup pathway. We have developed a novel biochemical assay which detects DNA strand invasion and exchange (key steps in HR), and we plan to use this assay to probe the importance of various nuclear proteins in HR. The assay was designed with dual-labeled complementary oligonucleotides using a fluorophore and a black hole quencher (BHQ). Labeling the oligonucleotides in this way allows us to monitor which oligonucleotides are annealed to one strand or another. We will use this biochemical assay to identify the nuclear proteins that are both sufficient and essential to perform the RAD52-dependent HR pathway. To further verify the importance of the proteins highlighted with these experiments, we will analyze RAD51 focus formation, and how it is affected by small interfering RNA (siRNA) elimination of each candidate protein. The possible implications of this study's results include the establishment of a potential gold-standard, novel biochemical assay, and the development of new therapeutic strategies for BRCA2-mutant cancers.

References

[1] American Cancer Society, "About Breast Cancer", [Online]. Available: https://www.cancer.org/content/dam/CRC/PDF/Public/8577.00.pdf

[2] R. Roy, J. Chun, and S. N. Powell, "BRCA1 and BRCA2: different roles in a common pathway of genome protection," Nature reviews. Cancer, vol. 12, no. 1, p. 68, Jan. 2012, doi: 10.1038/NRC3181.

526e Modeling Cancer Dormancy and Recurrence with the Theory of Birth-Death Processes

Adeyinka Lesi, David Rumschitzki

Unfortunately, a significant number of patients treated for cancer suffer a recurrence of the disease. Although most recurrences occur within the first 5 years after treatment, 10-20% of patients suffer from late recurrences (more than 10 years after treatment), where relapse can occur even after decades. Our novel population balance model frames the mechanism for recurrence as a random walk with size-dependent parameters describing growth (mitosis), reduction (immunity and/or treatment), and metastasis generation and aims to gain new insights about the proposed recurrence mechanism.

We leverage the theory of birth-death processes to predict the probability of a recurrence for a tumor of any given initial size in a large population of tumors and, for those that recur, the expected dormancy time. We do the analogous calculation for the cure probability and expected cure time for tumors of any given initial size. Similar approaches have been applied to predict extinction times in population ecology models and the fate of cells in genetic drift models. We also illustrate how the sizes at which growth and reduction rates intersect are threshold values that determine the evolution of the distribution. When only one such crossover size exists, it allows for a small fraction of tumors not too far below the crossover size to escape immune surveillance. When multiple crossover sizes exist, it may allow a stable or growing population of metastases to arise that eventually promotes recurrences. Using the analytical approaches described in this study, we established a new framework for understanding how recurrences in cancer might occur.

For validation, we are using our own observations of melanoma growth and shrinkage in immunocompetent zebrafish to obtain a size-dependent clearance time distribution for tumors that are cured. Using measured size-dependent growth and reduction rates, our model enables the prediction of expected clearance times. In zebrafish melanoma, we have found only one crossover size, and it is far larger than most tumors that we observe. The model can predict what fraction of large tumors can escape beyond this crossover size and the expected time for this to happen. Understanding how slowly or quickly tumors of different sizes should respond to treatment as well as the likelihood and expected time of recurrence can have important ramifications in a clinical setting, e.g., when adjuvant therapy might yield little benefit and potentially, given host response issues, even be detrimental.

526g Machine-Learning Based Prediction of Membranolytic Peptides with Anticancer Activities

Atefe Alimirzaei, Christopher Kieslich, Hyeju Song

Cancer is the second leading cause of death worldwide. Recently, membranolytic anticancer peptides (ACPs) have received considerable attention for their ability to target and kill cancer cells. Identification of ACPs is costly and usually time consuming. Therefore, development of efficient computational methods is of a great importance to aid in the identification of potential ACP candidates. In the current study, we developed support vector machine (SVM) models to predict membranolytic anticancer activity given a peptide sequence. Oscillations in physiochemical properties in protein sequences have been shown to be predictive of protein structure and function, and in this work, we are taking advantage of these known periodicities to predict if a peptide has ACP activity given the amino acid sequence. To this end, Fourier transforms were applied to the property factor vectors to measure the amplitude of the physiochemical oscillations, which served as the features for our SVM models. Peptides targeting breast and lung cancer cells were collected from the CancerPPD database and converted into physiochemical vectors using 10 property factors for the 20 natural amino acids. Also, we combined these datasets to investigate the model performance for discriminating active and inactive anticancer peptides regardless of the type of cancer they are targeting. Using the datasets, cross-validation has been applied to train and tune the models based on multiple training and testing sets with the accuracies around %75 and %80 for the models predicting peptides targeting breast and lung cancer cells, respectively. Also, for the model trained on the combined dataset this value was almost %78. The minimum number of features required to maintain accuracies is approximately 100 to 150 features for all models. However, given that our ultimate goal is the design of more active peptides, several works have been done to improve our model in separation active and inactive ACPs. We performed classification for multiple classes to more precisely predict specific activities of peptides. Instead of training a model with just two classes, active vs inactive ACPs, we developed 3 binary classification models which separately classify highly active ACPs versus moderately active ACPs, highly active ACPs versus inactive ACPs and moderately active versus inactive ACPs. The prediction results of the three individual classifiers combined together to predict whether a proposed peptide is inactive, moderately active or highly active, since we ultimately aim to design highly active peptides. Furthermore, to try to improve our prediction accuracy, we incorporated other sets of physiochemical features and properties of amino acids from the literature into our models and applied the cross-validation process again. To further validate our models, we also compared our predictions with models based on those features and properties.

526h Caspase-Dependent HDAC4 Translocation Due to Microsecond Pulsed Electric Field (μsPEF) Exposure of Glioblastoma Cells

Zahra Safaei, Gary Thompson

Introduction

Electroporation-based technologies using microsecond pulsed electric field (μsPEF) exposures are established as laboratory and clinical tools that permeabilize cell membranes to induce a variety of desired bioeffects. When used for cancer treatment, PEF exposures specifically triggers cell death through caspase activation [1-4]. nsPEF exposure induces apoptosis in different cell types through intrinsic and extrinsic apoptosis pathways, which both lead to activation of different caspases [5].

Overexpression of the class IIa histone deacetylase 4 (HDAC4) has been reported for glioblastoma cells [6, 7]. Dephosphorylation of HDAC4 leads to import into the nucleus and easy access to specific transcription factors, enabling repression of genes. Nuclear accumulation of HDAC4 promotes neuronal cell apoptosis [8]. In prior work, we have shown that μsPEF exposure can lead to nuclear accumulation of HDAC4 and a decrease in cell proliferation of U87-MG cells [9]. In this study we demonstrate the role of caspases in HDAC4 translocation under standard culture conditions and following μsPEF exposure of glioblastoma cells. These results can help to better understand the effect of μsPEF exposure on mechanisms of epigenetic modification.

Materials andamp; Methods

U87-MG cells were grown in T-75 flasks containing DMEM supplemented with 10 vol% fetal bovine serum (FBS, HyClone, SH30396.03, MA, USA) and 1 vol% antibiotic/antimycotic. To track HDAC4 localization, immunofluorescence assay (IFA) was performed 3 h after exposure using a HDAC4 polyclonal antibody (BioVision # 3604A-100, Milpitas, CA, USA) and a goat anti-rabbit IgG (Handamp;L) (DyLight® 488, NC, USA) as the secondary antibody and fluorescence marker.

Two kinds of custom buffer solutions were used throughout the experiments. Standard Outside Solution (SOS) consisted of 5 mM KCl, 2 mM $CaCl_2$, 10 mM 4-(2-hydroxyethyl)-1-piperazineethanesulfonic acid (HEPES), 2 mM $MgCl_2$, 10 mM Glucose, and 135 mM NaCl. Calcium-Free Standard Outside Solution (CAF) consisted of all the SOS components, except instead of $CaCl_2$, it contained 2 mM potassium ethylene glycol tetraacetic acid (K-EGTA). Solution pH was adjusted to 7.4 using NaOH.

A commercial electroporator (BTX Gemini X2 with PetriPulser electrode array and cuvette) was used to control square-wave, monopolar μsPEF duration (100 μs) and magnitude (1.45 kV/cm), while varying the number of pulses (0, 1, 10, 20 pulses, delivered at 1 Hz) to determine thresholds for HDAC4 translocation and proliferation reduction of U87-MG cells. For IFA, cells were cultured and exposed on glass-bottom petri dishes.

To identify the cause of cell death with respect to HDAC4 translocation and long-term effects of μsPEF exposure on HDAC4 localization, U87-MG cells were exposed to 10 consecutive pulses of 100 μs duration and 1.45 kV/cm at a repetition rate of 1 Hz. Then subcellular location of HDAC4 was acquired after 3h. To identify the role of caspase in HDAC4 translocation due to μsPEF exposure, caspases were inhibited by 50 μM Z-VAD-FMK (AAT Bioquest, 13300, CA, USA) with 1 h incubation. Each sample petri dish was filled with 1 mL of SOS or CAF containing inhibitor before pulse treatment. Then, cells were exposed to μsPEF. Samples were fixed with paraformaldehyde 3h after exposure, and the IFA protocol was performed as described above.

Results

Cells treated with the caspase inhibitor alone show significantly higher nuclear HDAC4 when compared to the untreated controls in CAF. The presence of extracellular Ca^{2+} neutralizes this effect, showing no significant changes between inhibitor-treated samples and controls (Fig. 1A). To determine whether caspases play a role in µsPEF exposure-induced HDAC4 nucleocytoplasmic shuttling, sham exposure controls are compared to 10 pulse exposures (Fig. 1B). The µsPEF exposure itself leads to nuclear import of HDAC4 in CAF. Caspase inhibition plus µsPEF exposure amplifies this trend in CAF. However, the presence of extracellular Ca^{2+} again does not show significant changes between shams and µsPEF exposures in SOS (Fig.1B). Comparing all samples receiving µsPEF exposure emphasizes the correlation between caspase activity and HDAC4 translocation (Fig. 1C). These results suggest that caspases at least partially determine nucleocytoplasmic shuttling of HDAC4 in CAF, whereas Ca^{2+} influx eliminates this effect by preventing nuclear accumulation of HDAC4.

Fig 1. The role of caspase activity in msPEF exposure-induced HDAC4 nucleocytoplasmic translocation in U87-MG cells is tested by pharmacological inhibition and shown by comparison of, A: Sham exposure controls, B: Sham controls vs. msPEF exposure, and C: msPEF exposures in solutions with and without extracellular Ca^{2+}. Statistical significance is tested using two-way ANOVA with Tukey's method, with symbols representing no significance (ns), P £ 0.0021 (**), P £ 0.0002 (***), and P £ 0.0001 (****).

References

1. Beebe, S. J.; Fox, P. M.; Rec, L. J.; Somers, K.; Stark, R. H.; Schoenbach, K. H., Nanosecond pulsed electric field (nsPEF) effects on cells and tissues: apoptosis induction and tumor growth inhibition. *IEEE Trans Plasma Sci* **2002,** *30* (1), 286-292.
2. Brock, R. M.; Beitel-White, N.; Davalos, R. V.; Allen, I. C., Starting a Fire Without Flame: The Induction of Cell Death and Inflammation in Electroporation-Based Tumor Ablation Strategies. *Front Oncol* **2020,** *10*, 1235.
3. Li, J.; Yuan, J., Caspases in apoptosis and beyond. *Oncogene* **2008,** *27* (48), 6194-6206.
4. Mortezaee, K.; Najafi, M.; Farhood, B.; Ahmadi, A.; Potes, Y.; Shabeeb, D.; Musa, A. E., Modulation of apoptosis by melatonin for improving cancer treatment efficiency: An updated review. *Life Sci* **2019,** *228*, 228-241.
5. Ren, W.; Beebe, S. J., An apoptosis targeted stimulus with nanosecond pulsed electric fields (nsPEFs) in E4 squamous cell carcinoma. *Apoptosis* **2011,** *16* (4), 382-93.
6. Lee, P.; Murphy, B.; Miller, R.; Menon, V.; Banik, N. L.; Giglio, P.; Lindhorst, S. M.; Varma, A. K.; Vandergrift, W. A., 3rd; Patel, S. J.; Das, A., Mechanisms and clinical significance of histone deacetylase inhibitors: epigenetic glioblastoma therapy. *Anticancer Res* **2015,** *35* (2), 615-25.
7. Reddy, R. G.; Bhat, U. A.; Chakravarty, S.; Kumar, A., Advances in histone deacetylase inhibitors in targeting glioblastoma stem cells. *Cancer Chemother Pharmacol* **2020,** *86* (2), 165-179.
8. Bolger, T. A.; Yao, T.-P., Intracellular Trafficking of Histone Deacetylase 4 Regulates Neuronal Cell Death. *J Neurosci* **2005,** *25* (41), 9544-9553.
9. Zahra Safaei, Gary Thompson. Glioblastoma treatment using epigenetic modification induced by microsecond pulsed electric field (µsPEF) exposure [abstract]. In: *Proceedings of the 113th Annual Meeting of the American Association for Cancer Research*; 2022 April 8-13; New Orleans LA. Philadelphia (PA): AACR; **2022**. Abstract nr 3314.

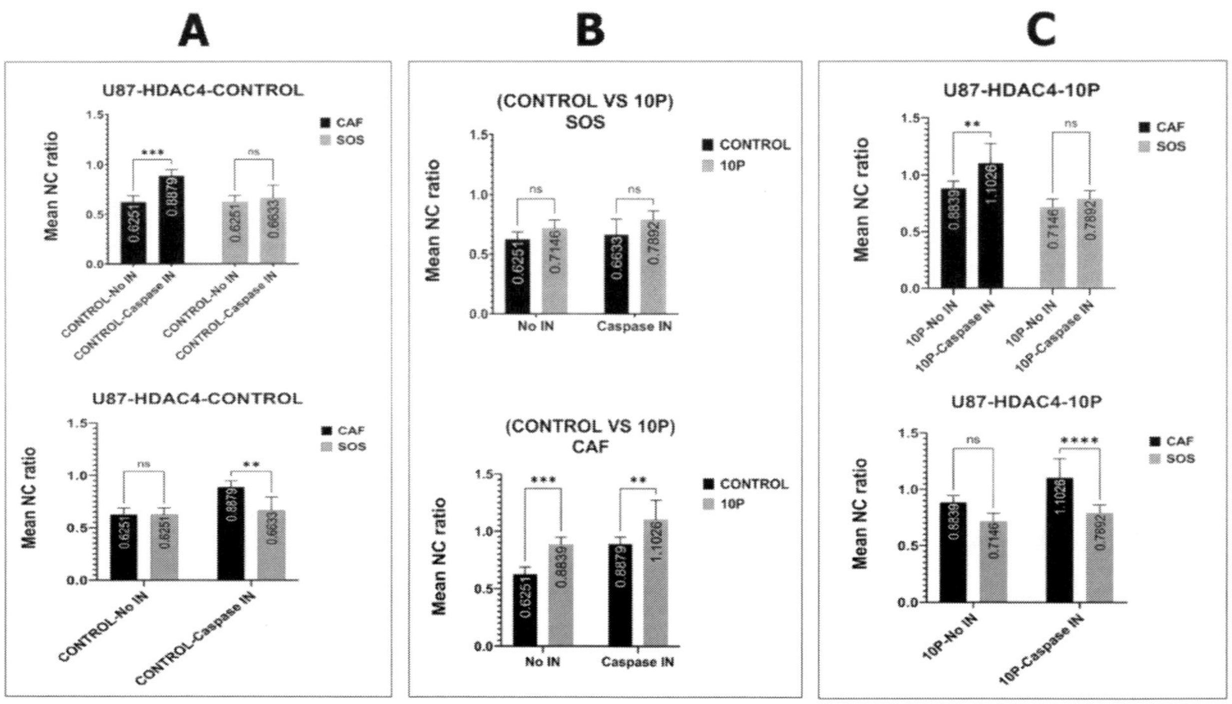
A
B
C
U87-HDAC4-CONTROL
Mean NC ratio
CAF
SOS
***
ns
0.6251
0.8879
0.6251
0.6633
CONTROL-No IN
CONTROL-Caspase IN
U87-HDAC4-CONTROL
ns
**
CONTROL-No IN
CONTROL-Caspase IN
(CONTROL VS 10P)
SOS
CONTROL
10P
0.6251
0.7146
0.6633
0.7892
No IN
Caspase IN
(CONTROL VS 10P)
CAF
***
**
0.6251
0.8839
0.8879
1.1026
U87-HDAC4-10P
0.8839
1.1026
0.7146
0.7892
10P-No IN
10P-Caspase IN
U87-HDAC4-10P
****

471a Mitochondrial Mechanisms Underlying NANOG Induced Reversal of Aging

Debanik Choudhury, Na Rong, Nika Rajabian, Georgios Tseropoulos, Pihu Mehrotra, Ramkumar Thiyagarajan, Pedro Lei, Izuagie Ikhapoh, Kenneth Seldeen, Bruce Troen, Stelios Andreadis

Mitochondrial dysfunction, one of the major hallmarks of aging has been implicated in age-related loss of stem cell function and development of inflammatory phenotype. Several cellular senescence models have shown deterioration of oxidative phosphorylation (OXPHOS), with senescent cells manifesting severe defects in respiratory chain complexes. In our study, we discovered that ectopic expression of pluripotent transcription factor, NANOG completely restored the mitochondrial function in senescent stem cells and that NANOG-driven rejuvenation led to metabolic reprogramming of senesced cells to the state of young proliferating cells.

To study the effect of NANOG on aged cells, we employed several models of aging i.e., replicative senescence of human Mesenchymal Stem Cells (MSC), myofibroblasts derived from patients suffering from Hutchinson's Gilford Progeria Syndrome (HGPS) and progeric mice model (LAKI). Young MSCs were induced to senescence by serial passaging. Cells were transduced with tetracycline-regulatable lentivirus, which enabled NANOG expression by addition of doxycycline, after cells reached senescence.

Senescent or aged cells exhibited severe mitochondrial dysfunction, as was evidenced by significantly reduced mitochondrial membrane potential, reduced respiratory function and excessive accumulation of reactive oxygen species (ROS). To delineate cell metabolic profile, we employed Seahorse analyzer and observed complete restoration of mitochondrial respiratory capacity in aged MSCs after NANOG overexpression. Metabolomics study revealed increased dependance of senescent cells on glutamine to meet their bioenergetic demand and this was confirmed by increased glutaminase (GLS) activity in aged cells. This observation also held true for our in-vivo study, where we observed increased GLS expression and activity in heart, aorta and skin of progeric (LAKI) as well as geriatric mice. Enhanced deamination of glutamine with senescence results in accumulation of toxic end-product, urea leading to detrimental consequences for mitochondrial function and DNA damage. Interestingly, inhibition of GLS activity in senescent cells by CB-839 decreased urea accumulation and partially restored mitochondrial function, decreased ROS and DNA damage. Furthermore, we examined the effects of CB-839 on LAKI mice and observed decreased age-associated ROS and urea accumulation in heart, skin and muscle, concomitant with improved mitochondrial function. Hence, inhibition of GLS activity rejuvenated mitochondrial function and led to amelioration of aging hallmarks.

In conclusion, our data provide novel insight into the mechanism, underlying metabolic reprogramming associated with cellular aging. Targeting glutamine metabolism and associated metabolites and byproducts may be a promising strategy to delay or reverse cellular senescence.

471b Effects of Processing Conditions on the Macroscopic Properties of Cellulose Filled Hydrogel Scaffolds Using UV Rheology

Bobby Haney, Subramanian Ramakrishnan

Hydrogel scaffolds are useful tools in observing and studying a variety of cells in synthetic complex environments. The anisotropic nature of some extracellular matrices require fabrication methods that permit structural alignment within the hydrogel. Extrusion based 3D printing of hydrogel scaffolds promotes shear induced alignment of anisotropic nanofillers or fibers and allows fast manufacturing and control via in situ polymerization. In this manufacturing method, shear at the walls of the nozzle align rod-shaped particles in the direction of extrusion, while ultraviolet light can be used to induce local crosslinking of the polymer solution to form a hydrogel. However, modulation of the final hydrogel scaffold by using this extrusion-based printing requires knowledge of the viscoelastic and optical properties of the polymer solution during the printing process. Also, to encapsulate cells in a 3D printed hydrogel at a high viability via "ink" extrusion and fast photopolymerization, optimization of the photo initiator concentration and light irradiation time is vital. In this work, a rheometer with an ultraviolet-light lamp attached is used to characterize the effects of photo initiator amount, light irradiation time, and initial polymer modulus on the final modulus of a crosslinked biocompatible hydrogel. Using alginate, photocurable polyethylene diacrylate, and high-aspect-ratio cellulose nanocrystals we characterize the crosslinking of a hydrogel when it is exposed to ultraviolet light to form a scaffold that's representative of the extracellular matrix. Due to the simultaneous crosslinking and light attenuation within the hydrogel, there exists an optimum photo initiator amount and irradiation time to control the final modulus. These settings are used to assess the final properties of a hydrogel formed immediately after experiencing the high concomitant shear forces of extrusion printing. It is found that the presence of long anisotropic fillers affects polymer solution transmittance and viscoelasticity upon shearing which leads to major changes in the final modulus of the scaffold.

471c 3D Bioprinting of iPSC Derived Islet Organoids in Hydrogel Constructs

Miranda Poklar, Ben Mizerak, Ravi Krishnamurthy, Connor Wiegand, Prashant Kumta, Ipsita Banerjee

One of the largest medical problems in the world is the lack of organs available for transplantation. In the US alone, as of February 2021, over 107,000 people were waiting for a transplant, and the organ need increases while the supply remains stagnant(1). 3D bioprinting is designed to fight this issue by printing biomaterial layers embedded with cells that can work to recreate a tissue construct. Primarily bioprinting has focused on constructing hard tissues, with less work going towards supplementing a soft tissue environment(2). Soft tissue requires bioinks that can act as support structures, while maintaining the correct biophysical properties for human cells. Another technology used to address organ shortages is induced pluripotent stem cells (iPSCs), as they can be differentiated to almost any tissue and tailored to patients. The current aim of this project is combining the advantages of bioprinting and iPSC organoids to form a complex soft tissue environment. In this study, iPSC islets were bioprinted with the goal of engineering endocrine pancreas tissue that can remain viable and retain functionality.

The ideal bioink was determined by optimizing the ink composition to produce a structure that provided the desired stability and compatibility with iPSC culture. A combination of 3% w/v alginate, and 6% w/v methylcellulose crosslinked with 50 mM CaCl2 was able to achieve these attributes, which was first confirmed with primary human islets. Cadaveric human islets (Prodo lab) were bioprinted in the ink and maintained in culture for over 3 days. The islets remained viable within the strand and exhibited the appropriate islet markers, c-peptide and glucagon as confirmed by immunofluorescent imaging. Islet function was confirmed by Glucose Sensitive Insulin Secretion (GSIS) assay, where the printed islet response was found comparable to control islets, with a stimulation index (Insulin at high glucose/ Insulin at low glucose) of 20.

Having confirmed retention of primary islet function in the bioink and printing configuration, we next adapted the procedure for iPSCs and iPSC-derived islet organoids. As a first step, we printed undifferentiated iPSCs which were suspended in the bioink as single cell (SC). The SC iPSCs self-assembled to form small aggregates during post-print culture within the print strand. The aggregates retained high viability over a 9-day culture, however, the process of self-aggregation resulted in cell debris within the strands, likely from the cells which failed to aggregate. As an alternative, we pre-aggregated the undifferentiated iPSCs in agarose microwells, and printed at an average size of 350 μm and cultured for 3 days. LiveDead imaging showed the aggregates maintained viability, while fluorescent staining showed the presence of Oct4 and Nanog, confirming pluripotency. Importantly printing of preformed aggregates significantly reduced the observed debris while printing single cells.

We next advanced with printing iPSC derived islet organoids. These organoids were derived using an organoid culture method established previously, where the islets exhibit pancreatic and islet markers, c-peptide and glucagon(3). iPSC islet organoids thus derived were suspended in the bioink and printed using the same procedure and maintained in culture for 7 days. LiveDead imaging showed the iPSC islets sustaining viability, while fluorescent staining indicated that they also retained expression of the pancreatic markers, NKX6.1 and PDX1, and islet markers c-peptide and glucagon. Work is currently ongoing to see if iPSC aggregates can undergo differentiation post-printing, and if after printing, iPSC islets can demonstrate proper islet function when observed under GSIS testing conditions. This project acts as proof of concept for bioprinting a functional pancreatic environment using iPSC islets, with future work focused on organ transplantation.

(1)"Organ Donation Statistics." OrganDonor.gov, Health Resources and Services Administration, 25 Feb. 2021, www.organdonor.gov/statistics-stories/statistics.html.

(2)Mandrycky, Christian, et al. '3D bioprinting for engineering complex tissues.' Biotechnology advances 34.4 (2016): 422-434.

(3)Richardson, Thomas et al. "Alginate encapsulation of human embryonic stem cells to enhance directed differentiation to pancreatic islet-like cells." Tissue engineering. Part A vol. 20,23-24 (2014): 3198-211.

471d Engineering the Skeletal Muscle for Improved Innervation after Peripheral Nerve Injury

Pihu Mehrotra, Shahryar Shahini, Nika Rajabian, Yali Zhang, Jianmin Wang, Song Liu, James Jablonski, Susan Udin, Stelios Andreadis, Kirkwood Personius

Peripheral Nerve Injuries (PNI) affect more than 20 million Americans and severely impact the quality of life by causing long-term disability. These injuries disproportionately affect young individuals and soldiers who fail to regain normal movement and demonstrate persistent weakness. The onset of PNI is characterized by nerve degeneration distal to the nerve injury resulting in long periods of skeletal muscle denervation. Since the speed of axon regeneration post injury is very slow (1-2mm/day), the skeletal muscle becomes atrophied by the time the nerves regenerate and is incapable of "accepting" innervation. To this end, we hypothesize that reprogramming the skeletal muscle to an embryonic-like state may preserve its reinnervation capability following PNI. Previous work in our lab had demonstrated that the transcription factor NANOG can rejuvenate skeletal muscle by increasing the number of $Pax7^+$ muscle progenitors and causes formation of embryonic myosin heavy chain (eMYHC) positive immature muscle fibers in vivo.

To study the effect of NANOG on skeletal muscle innervation, we generated a mouse model (Col1a1-tetO-Nanog, in the ROSA26 locus, hereafter referred to as ROSA-NANOG) in which NANOG expression can be induced upon exposure to doxycycline (Dox). To limit NANOG expression only in the muscle, a slow-release polymer (Elvax), was implanted subcutaneously near the tibialis anterior (TA) muscle for 2 weeks. Elvax implantation was accompanied by sciatic nerve transection and end-to-end repair in both ROSA-NANOG and wild type (WT). WT served as controls. Needle electromyography (EMG) recordings demonstrated that ROSA-NANOG mice elicited almost 3x increase in EMG amplitude as compared to WT mice 16wks post nerve injury, confirming improved reinnervation is a result of re-engineering the muscle to a permissible state. This functional recovery in muscle activity on nerve stimulation accompanied significant improvements in the toe-spread reflex of ROSA-NANOG mice at both 5wks and 16wks post injury. Interestingly, we found that NANOG expression in the muscle prevented atrophy and resulted in increased isometric force production as compared to wild type animals. Further, we performed whole muscle immunohistochemistry to quantify the overlap of muscular acetylcholine receptors (AChRs) with neurofilament and synaptic vesicles. Indeed, while there was very limited presynaptic overlap in WT muscle, ROSA-NANOG mice demonstrated extensive overlap between synaptic vesicles and AChRs indicating restored innervation. In conclusion, we demonstrate that re-engineering the muscle by ectopic NANOG expression can improve functional outcomes after PNI.

471e Regeneration of the Epithelium Layer Using Fibrous Biomimetic Basement Membranes in a Mouse Dermal Wound Model

Dina Gadalla, Maeve Kennedy, David Lott

Introduction: Due to the dynamic environment of the airway system, methods to restore respiratory epithelium during airway reconstruction are not suitable for large defect areas. A lack of epithelium in hollow organs such as trachea can lead to stenosis, collapse and infection that restrict airway patency. Consequently, engineered artificial tissue capable of regenerating a functional epithelium is required. This can be achieved by providing basement membrane biomimetic surfaces that are favorable to the migration and polarization of epithelial cells. To this end, we envision basement membrane adsorbed fiber surfaces – incorporated to the luminal side of tracheal replacements – could recruit host epithelial cells and stimulate epithelium formation at the interface between replacements and the environment. We sought to fabricate thin fiber layers that could adsorb a protein mixture to resemble both physical and chemical properties of basements membranes. Fiber layers were prepared by electrospinning, and a mixture of laminin and collagen IV was adsorbed to the surface thereafter. Response to the implant was studied *in vitro* and in an *in vivo* dermal wound mouse model.

Materials and Methods: Fiber layers (~100 μm thick) were prepared by electrospinning a 10% 99/1 wt/wt polycaprolactone (PCL)/heparin blend for 15 min. A protein mixture consisting of 10 μg/mL laminin and 20 μg/mL collagen IV was adsorbed to the fibers by incubating at 37^0C overnight. Scanning electron microscopy (SEM) was carried out to characterize fiber morphology and measure fiber diameters in the presence and absence of adsorbed proteins. The attachment of tracheal epithelial cells to the fibers was analyzed with a live/dead staining assay 1 and 3 days following cell seeding. Similarly, the migration of cells once in the vicinity of fibers was evaluated over a 5 day period both with qualitative staining and quantitative DNA content measurements. The presence of doses of adsorbed laminin and collagen IV proteins on fibers was characterized for its ability to induce the differentiation of basal cells to mature epithelium by establishing an air-liquid interface (ALI) culture. Gene expression of basal, mucous-secreting and ciliated cells were analyzed using NanoString® technology. Immunofluorescent staining was carried out to detect the presence of epithelium specific cell markers and SEM carried out to examine and confirm both cell presence and morphology. Next, a humanized dermal wound was developed to test the implants in mice. Here, 6 mm diameter implants were placed after creating 5 mm skin punches in the backs of mice. Wound photographs were taken every other day and the wound area measured each time. Additionally, immunofluorescent stainings to detect the presence of angiogenic and epithelial markers were carried out three weeks post-implantation.

Results and Discussion: The adsorption of proteins on the fibers allows fibers to retain the same fibrillar morphology yet introduces some roughness to the fibers indicating protein deposition and the consequent appearance of larger fibers. Cell adhesion studies indicate the presence of a larger number of cells attached on fibers in the presence of adsorbed proteins. Similarly, migration studies show an increase in cell numbers migrating to fiber layers in the presence of proteins. This is vital since native basement membranes mainly act substrates for the adhesion and migration of epithelial cells. Cell differentiation experiments show that fibers enhance the expression of several mucous secreting and ciliated cell genes. Staining indicates the presence of keratin and tight cell-cell junctions characteristic to epithelial tissue while the proteins on fibers stimulate the production of ciliated cells as implied with α-tubulin staining shown in Figure 1a. This was confirmed with SEM where tight mucosal globules with protrusions and microvilli that spread across the fiber layer were detected. Wound areas decreased over the study time period and an epithelium layer was regenerated over the surface of the implants as shown in Figure 1b. Furthermore, staining studies indicated the presence of CD31, α-SMA, pan-cyto keratin and E-cadherin in the wound area further confirming healing and wound area closure.

Conclusions: The presentation of basement membrane proteins using fiber layers increases cell adherence and cell migration to fibers, stimulates the differentiation of basal cells into ciliated cells and promotes epithelialization in an *in vivo* dermal wound model by mimicking the composition of native membranes. Thus, the incorporation of such protein adsorbed fibers on the luminal side of tracheal replacements could ultimately facilitate the regeneration of epithelial tissue and is the current next step we are exploring.

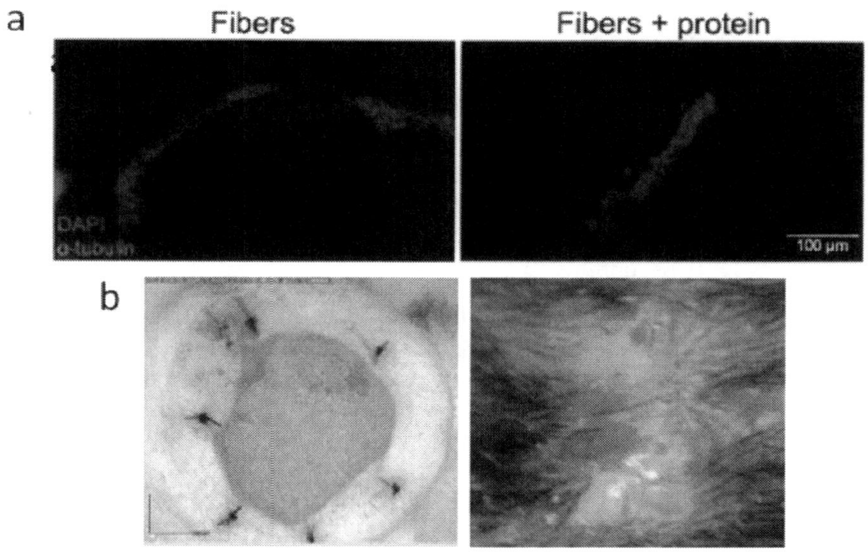

471f Photochemical Reaction Kinetics and Thermodynamics of Light-Induced Collagen Cross-Linking with Rose Bengal for Suture-Less Wound and Incision Closure and Repair

Alan Aguirre-Soto

Photochemical tissue bonding (PTB) of wounds and surgical incisions relies on the use of photosensitive molecules and light to induce the formation of covalent linkages in collagenous tissue. Rose Bengal (RB) has been found to be a suitable photosensitizer (dye) for PTB as well as for Photochemical Tissue Passivation (PTP). The light-induced cross-linking of collagenous matrices in the presence of RB has yielded promising results in laboratory and clinical trials in terms of wound closure, reduced inflammation and less scaring. However, the required irradiation doses are still higher than what would be ideal for broad implementation of this phototherapy. We recently proposed that the aggregation of RB molecules (monomers) lowers the quantum yield of reactive oxygen species (ROS) molecules. Kinetic steady-state analysis showed that dimeric species are not as effective as monomeric RB for the photochemical ROS-mediated therapy. Here, we tested several cucurbituril and cyclodextrin based nanocages (NCs) as potential inhibitors of RB aggregation to study the effect of dimerization on RB photobleaching and oxygen consumption kinetics when RB is irradiated with green light in aqueous phosphate-buffered solutions.We discuss the effects of RB and nanocage concentrations in a proposed mechanism for the RB light-induced oxidative decomposition process that has previously been linked to the formation of crosslinks in the presence of collagenous tissue. Our proposed mechanism integrates thermodynamic and reaction kinetics of the RB photo-oxidation reaction via the monitoring of the photochemical reactions with multiple spectroscopic techniques in tandem. Notably, we utilized a fluorescence-based oxygen sensor to monitor the concentration of molecular oxygen in real-time. We confirmed that the RB/HP-γ-CD host-guest complex yields the fastest photobleaching and oxygen consumption rates. The cyclodextrins appear to be better candidates for precluding RB aggregation, making them great candidates for novel PTB approaches with higher efficiencies and lower fluence requirements. We provide a comprehensive analysis of the photochemical and photophysical pathways of nanocaged RB to better understand how to control the function of RB as a photo-active ROS-therapeutic agent. We hope that this contribution helps advance the field of photomedicine in general, and the implementation of PTB for tissue bonding and repair in particular.

284a Data-Driven QSAR Modeling for the Iterative Identification of Chemical Motifs from Limited Data

Hyun-Seob Song, Chul Min Park, Soojin Jang, Jae Sun Kim, Hyung Chul Ryu, Hasoo Seong

Drug discovery and development is a lengthy and costly process composed of several complex stages, including target validation, high-throughput screening, hit to lead, lead optimization, preclinical/clinical tests, and post-clinical evaluation. Advanced data-driven modeling techniques such as artificial intelligence and deep learning (DL) offer the promise to revolutionize the drug development pipeline through significant reduction in time and cost. One of the key utilities of DL models is to build QSAR (quantitative structure–activity relationship) or QSPR (quantitative structure-property relationship) models that enable accurate identification of the relationships between chemical structures and their activities or properties by leveraging big data. While groundbreaking, development of predictive DL models is often challenging, particularly when sufficient datasets required for training are not available, a common issue encountered at an initial stage of research.

In this work, we present a novel workflow to address this limitation. Our strategy is to develop a relatively simple data-driven model using a small set of data and subsequently use this initial model to guide the next round of experiments for generating new datasets. The resulting data is then fed back to the model for refinement. We continue this iterative process until the predictions of QSAR or QSPR models are consistent with experimental measurements. During this process, the complexity and depth of the model are accordingly increased. Once validated, the model is finally used to suggest new chemical structures that potentially outperform any of the previously tested chemical compounds. We demonstrate our pipeline through the case study of *Staphylococcus aureus*, a pathogen causing various life-threatening infections, including skin, lung, heart valve and bone infections. We represent structural information of compounds using the SMILES (Simplified Molecular Input Line Entry System) code; define atomic and symbolic features for each of the entries in the SMILES string; develop a first version of the model using SINDy (Sparse Identification of Nonlinear Dynamics). Despite data deficiency, this initial model shows reasonably accuracy and enables identifying chemical motifs (i.e., substructures of compounds) that may determine the drug activity. Furthermore, the model allows us to evaluate a comprehensive list of compounds in public databases and select top candidates based on the model-estimated activities. The workflow developed through this work therefore greatly facilitates otherwise time-consuming steps in identifying drug candidates through iterative interactions between modeling and experimental groups.

284b Single Sequence Prediction of Protein Structure and Impacts on Computational Protein Design

Ratul Chowdhury

The long-term goal of Chowdhury Lab is to simulate in a bottom-up fashion, a simple metazoan cell with sufficient fidelity such that any experimentally measurable cell-biological quantity can be predicted *in silico* with comparable accuracy. To this end we have first set up a deep learning pipeline to predict protein structures directly from sequences using natural language processing. This paves the path for us to understand protein-protein and protein-non-protein interactions. These models extract the 'grammar' of protein folding from a corpus of all known ~250M protein sequences and unravel general rules that guide the folding patterns. Much like the real biological process of protein folding, which does not depend on any sequence alignment, we outperform MSA-dependent methods like AlphaFold2 on certain protein classes. We predict protein structure from its amino acid sequence without evolutionary data (in the form of MSAs). We thus map protein sequences to high-dimensional spaces to cluster functionally proximal proteins as nearby points without needing expensive 'labelled' data (e.g., experimental PDB structures).

We explain how this lays the foundation to construct, biochemistry-aware, machine learning models of proteins and their interactions to (a) derive insight about their functions, (b) tune existing proteins for biochemical and pharmaceutical applications, and (c) point-of-care diagnostics.

284c Hierarchical Graph-Based Representation Drives Prediction of Stapled Peptide Drug-like Properties

Marshall Case, Camille Bilodeau, Greg Thurber

Sequence-based models like AlphaFold and UniRep have led to transformational improvements in globular protein structure prediction and quantitative function prediction, respectively. However, there are a wealth of applications where proteins based on the 20 canonical amino acids are insufficient to address biomedical, biosensing, and biotherapeutic challenges. These non-standard modifications often include modifications beyond the linear information contained in protein sequences, such as intramolecular disulfide bonds, which are difficult to incorporate in sequence-based models. One such application is stapled peptides, chemically modified peptides that overcome pharmacokinetic barriers of the human body necessary to serve as efficacious therapeutics; these barriers include maintaining structural integrity through high proteolytic activity, penetrating the highly hydrophobic cellular membrane, and having the affinity and specificity to efficiently engage disease related proteins. To overcome these challenges, a "stapled" peptide can be formed by covalently linking two amino acids, locking the peptide into its alpha-helical state. Recently, the Thurber Lab has pioneered a technique known as **S**tabilized **P**eptide **E**ngineering by ***E****. coli* **D**isplay (SPEED) to rapidly characterize and develop $>10^9$ unique stapled peptide therapeutics, minimizing the need for expensive and low-throughput solid phase synthesis. However, the process of narrowing the selection of peptides from 10^9 members to a small handful for *in vitro* or *in vivo* analysis is non-trivial.

In this work, we design and validate a hierarchical, graph-based model to predict and optimize properties of stapled peptides and apply it to identify lead compounds for translation to *in vitro* models. Because the complex topology of stapled peptides is challenging to represent, we employed a message passing graph neural network (MP-GNN) to produce a machine interpretable, fully differentiable representation. First, canonical and non-canonical amino acids alike are represented as vectors, instantiated with quantum chemical descriptors like charge, hydrophobicity, and bond order. Then, information about local chemical environments is encoded by passing messages between atoms, updating the vector representations at each step. Next, because a peptide's properties are a function of its amino acids, each amino acid is represented as a sum of its atom vectors and the same message passing process is repeated at the amino acid level. This encodes global information about the peptide, which theoretically includes information such as intramolecular NH_3^+ - $COOH^-$ salt bridges between charged amino acids and π-π interactions between aromatic amino acids.

We demonstrate the power of this model by designing stapled peptides towards the Bcl-2 proteins, or B cell lymphoma 2 proteins, which regulate apoptosis within cells and are often overexpressed in cancer cells. Because these proteins are highly related in sequence but play distinct roles in apoptosis regulation, methods and models that optimize specificity are in great demand. We designed a library of stapled peptides that inhibit Bcl-2 proteins and therefore induce cancer tells to undergo apoptosis, screened it for desired properties using the bacterial cell surface, and generated quantitative property labels using Next Generation Sequencing. These labels are combined with the unsupervised representation from the MP-GNN of stapled peptides to train and validate the model. Using this model, we can explore the sequence-function space of stapled peptides. Finally, lead compounds are identified by optimizing peptides to the frontier of idealized target properties. Importantly, the peptides identified through modeling had improved affinity and specificity compared to ones identified directly through sorting. The ability to improve affinity, specificity, and stability of stapled peptides demonstrates the value of a model that can handle proteins with geometry beyond naturally occurring proteins. This model could eventually lead to better prediction and generation of similar molecules, such as cyclic peptides or glycoproteins.

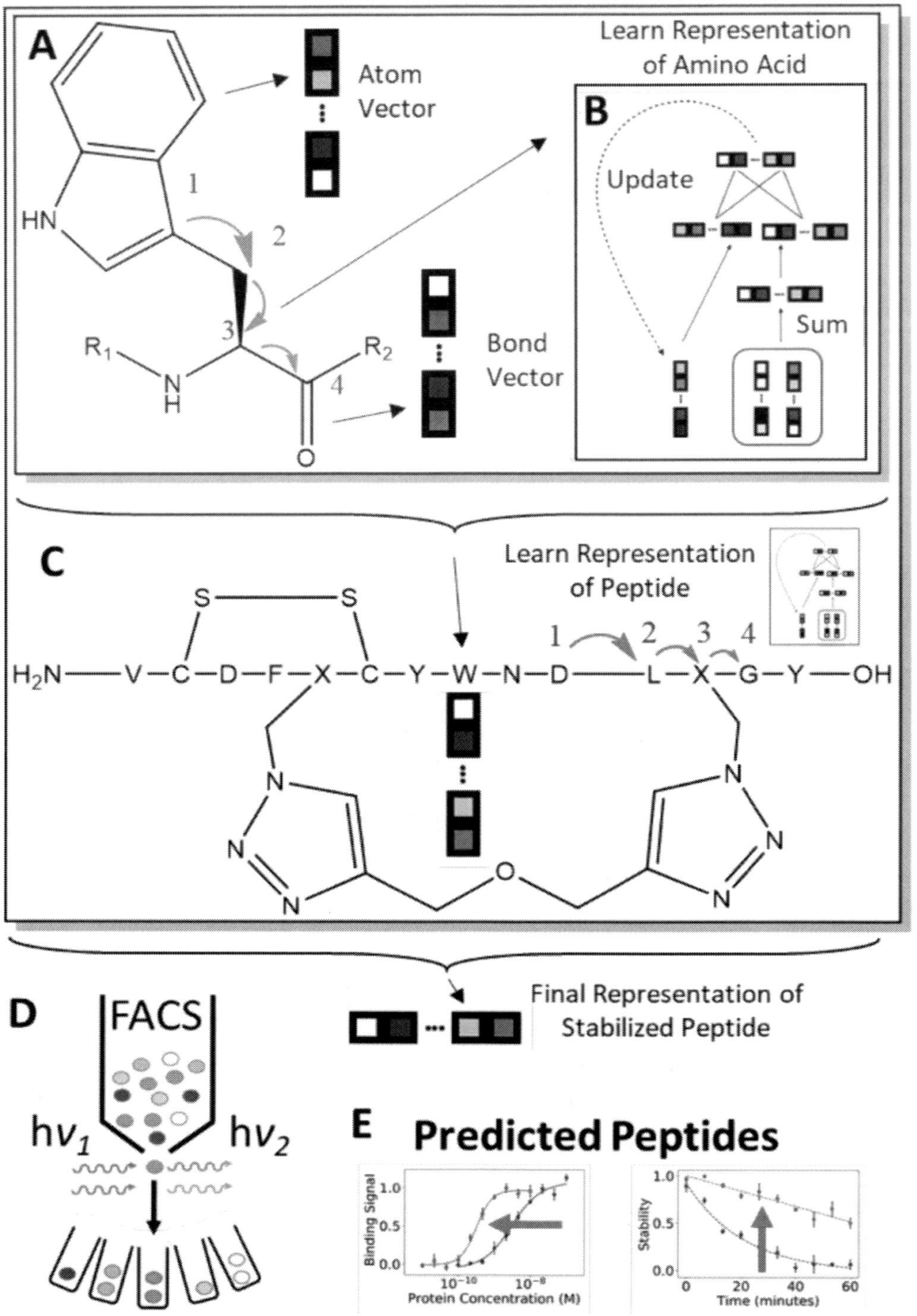
A
Atom Vector
Learn Representation of Amino Acid
B
Update
Sum
Bond Vector
C
Learn Representation of Peptide
H2N—V—C—D—F—X—C—Y—W—N—D—L—X—G—Y—OH
Final Representation of Stabilized Peptide
D
FACS
hv1
hv2
E
Predicted Peptides
Binding Signal
Protein Concentration (M)
Stability
Time (minutes)

Figure 1: Prediction of stapled peptide properties using message passing graph neural networks. **(A)** First, atoms are represented as vectors of quantum chemical information. **(B)** Local chemical environments are encoded by passing messages between atoms and updating vector representations at each step. **(C)** Then, each amino acid is encoded as a sum of its atom vectors and the same message passing process is repeated at the amino acid level. **(D)** Quantitative labels are obtained from fluorescent activated cell sorting experiments using surface-displayed stapled peptides. **(E)** Finally, unsupervised peptide vector representations and function labels are fed into a neural network to predict and improve peptide functions, such as affinity (left) and stability (right).

284d Data-Driven Approach for the Prediction of MHC Class II Epitopes Using Oscillations of Physicochemical Properties

Hyeju Song, Christopher Kieslich

Identifying T-cell epitopes in antigens is of great interest to improve understanding of cellular immunity and to aid the design of peptide-based vaccines, therapeutics and diagnostics. Traditional epitope identification relies on costly and time-consuming experimental techniques, thereby computational epitope prediction methods have emerged for in silico screening of peptides with the increasing wealth of gene and protein sequence data generated by high-throughput technologies.

Major histocompatibility complex (MHC) class II molecules, expressed on the surface of antigen-presenting cells (APC), display peptides to be recognized by CD4+ T-cells which would elicit various host immune responses. Thus, binding of peptides derived from protein antigen to the MHC molecules is a prerequisite for T-cell immunogenicity. One approach for the computational prediction of peptide-MHC binding is the data-driven machine learning method which involves predicting binding affinities given the sequences of the peptide and an MHC molecule. Numerous prediction tools have been developed for peptides-MHC class II binding, but it remains a challenging problem because of the polymorphic nature of MHC class II molecules and the variations in peptides length.

The presented work tests the performance of support vector machine (SVM) models of multiple allele-specific models combined with a previously proposed SVM based feature selection algorithm. The SVM models aim to classify MHC class II binding and non-binding peptides based on their amino acid sequences and derived features. In developing the SVM model, we take advantage of underlying periodicities in physicochemical properties along the sequence of a peptide that have been shown to be predictive features. Once the physicochemical descriptors are generated, Fourier transforms are then applied to be able to encode peptide sequences of varying lengths. In training and testing the model, a comprehensive dataset of MHC class II binding peptides was taken from IEDB database and cross validation and grid search are applied across multiple train and test datasets. A feature selection algorithm is also incorporated into the model development to identify an essential set of predictive features.

284e Automated Detection of Apoptotic Bodies in Label-Free Time-Lapse High-Throughput Microscopy Using Deep Convolutional Neural Networks

Kwan-Ling Wu, Melisa Martinez-Paniagua, Kate Reichel, Prashant Menon, Shravani Deo, Badrinath Roysam, Navin Varadarajan

Immunotherapy has altered the cancer treatment landscape and enabled clinical responses lasting for decades. Most treatment modalities remove tumor cells by inducing programmed cell death, or apoptosis, which involves the release of apoptotic bodies (ApoBDs) from dying cells. ApoBDs can augment the durability of the immune response as they are efficiently phagocytosed and catalyze an adaptive immunity in the host. Therefore, a robust approach to identifying the induction time of apoptosis in time-lapse assays is necessary for a deeper mechanistic understanding of immunotherapy and further advances in therapeutics. The multiplexing limitations of high-throughput microscopy and the biochemical perturbations caused by fluorescent labels in conventional methods motivated us to investigate label-free detection methods. However, identifying ApoBDs in label-free images using traditional pattern recognition algorithms is challenging due to their small size, complex morphology and variability in size. We present a label-free phase-contrast microscopy based high-throughput method to detect emergences of ApoBDs in human melanoma cells using deep convolutional neural network (CNN) based image analysis with minimized human annotation effort. Reliable analysis of the visual appearance of ApoBDs was made possible by the ResNet50 classifier, providing an accuracy of 90% when detecting ApoBDs in single frames, and frame-based error < 1 frame (5 min) when estimating the ApoBD formation time in image sequences. To reduce the manual annotation labor for training the CNN, we developed a strategy that combines blub detection and gradient-based localization to generate annotations for training the CNNs. The Mask-RCNN trained with an image dataset augmented by a saliency heat map reached an intersection over union (IoU) accuracy of 75% when segmenting individual ApoBDs. Our results provide a robust tool for ApoBD detection and a resource-efficient way to generate strong annotation for image-based assays.

284f Single-Cell Transcriptomic Analysis of Neural Development

Karthik Shekhar

Background: Recent technological breakthroughs in single-cell genomics have facilitated the interrogation of complex biological tissues at unprecedented molecular detail. Here, the ability to marry high-throughput measurements with data-driven machine learning algorithms enable the inference of genetic rules that underlie not only the diversity of cell types present in tissue, but also dynamical processes such as differentiation and disease pathogenesis. Our group is applying these approaches to understand developmental questions in the brain. *A central question in neuroscience is how the brain's diverse neuronal types – defined by location, connectivity, function and molecular profile – are established via developmental processes.* The vertebrate retina has been a valuable model to address this question. Lineage tracing, where the fate of a mitotic progenitor cell is tracked using an indelible marker, has provided deep insights into when and how broad classes of neurons are generated. However, the diversification of a neuronal class into multiple closely related types happens postmitotically, and is inaccessible to classical lineage tracing methods.

Goals: To address this challenge, we have used single-cell RNA-sequencing to understand the postmitotic diversification of immature neuronal precursors in two parts of the visual system, the retina and the visual cortex. Here, the identity of a neuron is associated with a multi-dimensional molecular state (e.g. expression profiles of thousands of genes). By applying machine learning methods to tens of thousands of single-cell snapshots, we can not only identify molecular states that define cell types, but also infer their process of diversification and molecular maturation. I will discuss two examples in the retina that illustrate the power of this approach that bridge both basic and translational questions.

Results: First, I will describe the process by which retinal ganglion cells (RGCs), the sole output neurons of the retina, diversify into ~45 discrete transcriptomic types from immature postmitotic precursors. We find that the number of molecularly distinct groups of RGCs as well as their transcriptomic distinctiveness increases with age. By adapting a mathematical approach known as optimal transport, we obtained evidence that RGC types are not specified at the progenitor level but rather that multipotentiality persists at the postmitotic stage in precursor RGCs. These precursor RGCs become gradually and asynchronously restricted to specific types with developmental age. We identified temporally regulated modules of genes that correlate with, and likely regulate, multiple aspects of RGC development, ranging from differentiation and axon guidance to synaptic recognition and refinement. Diversification may in many cases occur in two steps, with precursors initially committing to subclasses, each defined by the selective expression of TFs in the adult. Subsequently, precursors within a subclass, become restricted to single types by a process we named "fate-decoupling".

Second, using these results as a foundation, we analyzed the impact of disrupting vision-dependent activity on RGC diversification. We analyzed RGC transcriptomes in mice lacking visual experience due either to dark rearing from birth or to mutations that ablate bipolar or photoreceptor cells. We find, perhaps unsurprisingly, that visual activity is dispensable for diversification as > 98.5% of visually deprived RGCs can be unequivocally assigned to one of 45 types based on their transcriptomes. However, the number and magnitude of type-specific differentially expressed genes were decreased in VD RGCs, implying that activity influences RGC maturation or maintenance. Consistent with this notion, transcriptomic alternations in VD RGCs share a significant overlap with gene modules found in developing RGCs. Together, our results provide a resource for mechanistic analyses of RGC differentiation and maturation, and for investigating the role of activity in these processes.

284g Structure-Guided Functional Phylogenetic Trees of Human Channel Proteins and Kinases for Drug Discovery

Ratul Chowdhury

Kinases mediate a large number of cellular processes (such as metabolism, proliferation, and apoptosis) via catalyzing protein phosphorylation. Kinases have thus emerged as promising drug targets for oncology. However, the large size of the kinome (~500 human kinases) and influence of activation states upon drug binding makes it difficult to discern their cellular modes of action. Clustering analyses across the human kinome (518 kinases) are limited to clustering them based only on sequence similarity - from a multiple sequence alignment (MSA). ML-based structure predictors like AlphaFold2 have supplemented experimental PDBs and thus allows us to access 3D-structures of the entire human kinome including the less understood dark kinome. Functionality of kinases is majorly guided by structural similarity and the presence of similar binding pockets for drug accessibility. We introduce a consolidated CLASSIF (Clustering Annotated Sequence, Structure, and Information of Function) workflow to: (a) predict/use kinase PDB structures, (b) annotate kinase domain geometries, (c) identify ATP binding domains (d) align ATP binding pockets in structure as well as latent space, (e) integrate experimental drug binding affinity with sequence and structural data, and (f) construct phylogenetic trees based on these sequence-structure-function metrics. We provide a computational framework to discern the biophysical bases of drug-kinome interactions at the atomic level and extend this on to dark kinases and novel drugs alike.

While kinases are shorter proteins (<400aa) and have nearly 75% of the kinome experimentally resolved, voltage-gated ion channels constitute a far more challenging set. These channels are often ~1500 aa long single-chain proteins, with barely 5% structures resolved experimentally. These neuron-bound surface proteins are therapeutic targets for pain medications. We challenge our deep-learning-based structure-prediction tool RGN2 to first predict the structures of all ~130 human voltage-gated ion channels. Preliminary results after passing these channel structures through the CLASSIF workflow have been able to recover conformational binding modalities of known pain killer drugs like lidocaine.

284h Machine Learning Enabled Cancer and Immune Cells Segmentation on Clinical Images

Vidit Shah, Felicia Poynter, Shachi Mittal

Motivation: Cancer Diagnosis typically involves a manual examination of stained tissue samples to capture different disease patterns. This task is often time-consuming, subjective, and limited in observing all the heterogeneous components. Manually utilizing pattern information from multiple cell types can be difficult, and computerized pattern recognition has been reported to be effective[1–4]. Additionally, our lack of understanding of the progression of this disease limits the ability to provide timely, holistic, and personalized care to the growing number of cancer patients around the world. There are several molecular and physiological changes that together govern the course of the disease. Currently, many immunohistochemical biomarkers are used to investigate breast tumors and determine their prognosis. Some of these biomarkers, like tumor-infiltrating immune cells, are important targets for new therapies. There is a lack of standardized assessment of tumor-infiltrating lymphocytes (TILs, an important class of immune cells) on stained tissue specimens, as suggested by the International Immuno-Oncology Biomarker Working Group[5,6]. Machine learning coupled with clinical images can provide a quantitative and objective platform for digitally segmenting cancer cells and immune cells for accurate diagnosis and comprehensive cancer profiling. Here, we present a machine learning based digital disease segmentation identifying different tissue components on breast cancer biopsy images.

Methods and Results: This work utilizes archival tissue samples from clinics and biobanks, primarily formalin-fixed and paraffin-embedded (FFPE). The samples are stained using Hematoxylin and Eosin (Handamp;E) staining, the most commonly used stain in medical diagnosis. These samples are imaged using a whole slide image scanner that provides a high-resolution image of the entire tissue sample. We have predominantly used datasets from public healthcare repositories in this work, but we will also extend our pipelines to datasets from our clinical collaborators. We use a combinatorial approach of unsupervised learning for data exploration to separate heterogeneous cell components and supervised learning to capture diagnostic classes in the sample. First, a simple deep learning model (previously developed by Mittal et al. [7]) is utilized to digitally segment the epithelial and stromal compartments on the Handamp;E stained image. The developed pipeline takes an Handamp;E image as an input and divides it into small regions of interest at different zoomed levels. The goal was to mimic the protocol followed by a pathologist for diagnostic decision-making. Next, these compartments are individually segmented using clustering and image segmentation to identify the different populations of the cells. This platform allows the identification of cancer cells and microenvironment sampling, particularly the immune cells, as illustrated in the figure below. The model determines the different tumor subpopulations and the corresponding immune cells in the tumor bed. In the stromal areas, the model captures various pockets of the collagen (altered density and polarization) and the intermixed cellular components like immune cells, fibroblasts, etc.

Outcome: The developed platform provides the ability to conduct quantitative, rapid, and standardized visualization of tumor subpopulations, stromal signatures, and immune cells, allowing clinicians to determine a comprehensive/accurate disease profile.

References:

1. Beck, A. H. *et al.* Systematic Analysis of Breast Cancer Morphology Uncovers Stromal Features Associated with Survival. *Science Translational Medicine* **3**, 108ra113-108ra113 (2011).
2. Janowczyk, A. andamp; Madabhushi, A. Deep learning for digital pathology image analysis: A comprehensive tutorial with selected use cases. *Journal of Pathology Informatics* **7**, 29 (2016).
3. Litjens, G. *et al.* A survey on deep learning in medical image analysis. *Medical Image Analysis* vol. 42 60–88 (2017).
4. Ehteshami Bejnordi, B. *et al.* Diagnostic Assessment of Deep Learning Algorithms for Detection of

Lymph Node Metastases in Women With Breast Cancer. *JAMA* **318**, 2199 (2017).
5. Kos, Z. *et al.* Pitfalls in assessing stromal tumor infiltrating lymphocytes (sTILs) in breast cancer. *npj Breast Cancer* **6**, 1–16 (2020).
6. Dieci, M. V. *et al.* Update on tumor-infiltrating lymphocytes (TILs) in breast cancer, including recommendations to assess TILs in residual disease after neoadjuvant therapy and in carcinoma in situ: A report of the International Immuno-Oncology Biomarker Working Group on Breast Cancer. *Seminars in Cancer Biology* vol. 52 16–25 (2018).
7. Mittal, S., Stoean, C., Kajdacsy-Balla, A. andamp; Bhargava, R. Digital Assessment of Stained Breast Tissue Images for Comprehensive Tumor and Microenvironment Analysis. *Frontiers in Bioengineering and Biotechnology* **7**, 246 (2019).

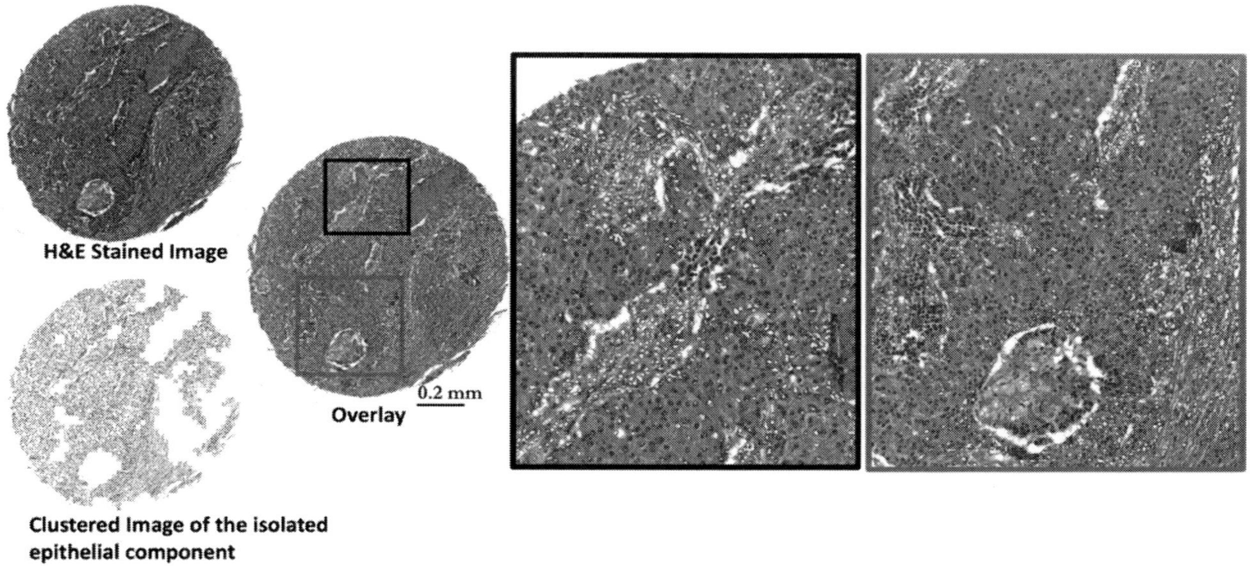

Fig.1: Clustering of the isolated epithelial component (using deep learning) on H&E stained image where each color corresponds to a heterogeneous cellular component

348a Indoor Transport of Aerosolized Respiratory Droplets in Multiroom Buildings

Alex Vlachokostas, Timothy Salsbury, Carolyn A. Burns, Leonard Pease

Here we describe our investigations into how virus laden respiratory droplets transport in multi-room buildings served by a centralized ventilation system. Even as many countries refine COVID-19 protocols, the pandemic and other respiratory diseases continue to persist worldwide. Most transmission has been measured and attributed to indoor air spread of virus laden respiratory droplets, yet most models of aerosolized droplet transport through buildings only consider one isolated room (without the interconnectivity of a modern air handling system) and do not take into account the chemical complexity of respiratory droplets (e.g., by assuming respiratory droplets are analogous to salt water droplets without mucus). This raises several important questions. How do air handling systems affect the spread of respiratory droplets? How do ventilation rates, air filtration, and outdoor air intake affect probabilities of infection in multiroom buildings? How can mucus affect the persistence of enveloped viruses that need to remain hydrated to maintain infectivity? What steps can be taken to minimize infectivity within buildings? This presentation addresses these important questions.

348b Multi-Strain Integrated Modelling for COVID-19

Dimosthenis Sarigiannis, Achilleas Karakoltzidis, Spyros Karakitsios

Pandemic evolution prediction has started to become particularly complex, as a result of the various parameters that affect the epidemiological course, including (a) the non-pharmacological interventions (such as social distancing, extensive testing and self-testing), (b) the pharmacological interventions (vaccines), (c) the effect of seasonality (d) the different transmissibility, disease severity and reinfection capacity of the various strains.

All the above parameters, have a critical effect on the course of the pandemic, rendering it practically impossible to successfully track it with the simple SIR and SEIR models. On the contrary, all of the above have to be described dynamically in time by more complex equilibriums, having just as a starting point the SEIR models. Accounting for all of the above, and having in mind the decision making support to the national health system over the COVID-19 pandemic, a multi-modal computational tool (called CORE: COVID Risk Evaluation model) for evaluating the COVID-19 epidemic health risk in Greece, Italy and USA has been developed, able to access the impact of various pharmacological and non-pharmacological interventions. CORE also describes in detail the various states of the disease severity by capturing reliably its dynamics, including the impact of risk reduction measures, as well as the impact of the new strains. The CORE model has already been included in the OECD COVID-19 computational toolbox. For defining the contact matrix among the various population groups, accounting for their sociodemographic profiles (i.e. age, occupation etc), as well as the impact of targeted social distancing measures an Agent-Based Modelling (ABM) (Chapizanis et al., 2021) has been employed.

Of particular interest in the latest months, is the introduction in the model of the 2 new dominant strains, i.e. Omicron (by the end of November 2021) and Omicron-2 (by February 2022). The greatest challenge in the modelling effort was the description of the susceptible population equilibrium, since this is affected by (a) the extent to which natural immunity from one strain ensures immunity of the other strain and (b) the differences in the immunity and the efficiency against infection of the people vaccinated with 2 and 3 doses and (c) the gradual loss of immunity, either natural and/or acquired. This was resolved by describing the full set of all the equilibriums of each strain (Δ, O, O-2) for the various disease states independently, interacting at the level of the susceptible equilibrium; different fraction of the ones that had acquired immunity by one strain, where also deducted from the available susceptible population of the other strains. At the same time, differences in the transmissibility among the strains, as well as the differences of re-infection capacity between strains were explaining very well the pandemic evolution in the number of overall infections on a daily basis. Similarly, the differences in the severity between Δ, O, O-2 strains have been accounted for, reflected in the successful prediction of the people that were going to require hospitalization in critical condition (intubated in ICU).

The evolution of the COVID-19 dynamics for Greece, as well as the long-term prediction till the end of June 2022, accounting for the current existing measures, vaccination and self-testing rate, is illustrated in Figures 1a (7d moving average of new infections per day) and Figure 1b (daily number of people intubated in ICU). The implementation of the Omicron strain in the beginning of the December in Greece was critical, because in this way we have been able to identify at a very early stage the already significant dispersion of this highly contagious strain and to warn timely the public authorities. In fact, the increase on the rate of change of the new daily cases by mid-December (Figure 1a) could not be explained by the current social distancing measures and the extent of population vaccination, but it could be very well explained only by accounting for a higher extent of Omicron dispersion in the community, than the one that had been identified by the sequencing of random samples. This allowed us to identify the timing and the height of the pandemics peak and the very rapid decline following that. However, the rapid decline was slowed down retarded after 10 days, as a result of the delay in the third dose of vaccination and the significant difference of the efficiency against infection between the second and the third dose.

Similarly, an early warning has been provided for the presence of Omicron 2 by the end of February, in contrast to the common belief that the pandemic was finally going to an end (Figure 1a).

The number of new infections observed nowadays, is significantly higher than the expected when the de-escalation of the 5th wave began (4th of January) and indicates loss of immunity. The main reason for this loss of immunity is the delay in the application of the booster dose, given the significant difference in the effectiveness of protection between vaccinated with 2 doses and with 3. Also, a significant loss of immunity exists due to the increased rate of re-infections even in short period of previous infection, which is characteristic of the new subtype of Omicron 2, which is assumed to be even more contagious than variant C by 33% mean (Lyngse et al., 2022). Therefore, the combination of delay of the 3rd dose and the presence of the new strain, led to a new wave has already further delayed the faster de-escalation.

In general, in the last 3 months, the theory of dynamic equilibrium that has already been formulated by our team has been confirmed. In this case, the relaxation of the social distancing measures by mid-February, allowed the room to the most contagious Omicron 2 mutation to disperse and to result in an additional peak (practically forming a 6th wave), further delaying the rapid de-escalation of the pandemic. However, a very high 6th wave due to Omicron 2 has been avoided by the additional natural immunity that has been obtained (and added to the overall acquired and already existing natural immunity), as a result of this fluctuation. An additional aggravating (and of course uncertainty) factor that has to be taken into account, is the significant number of re-infections (accounting for almost 10% of the number of daily cases at the moment), as well as the impact of reduced immunity in people who were vaccinated with the 3rd dose more than 4 months ago. Given the above, loosening existing restriction measures should be done with great attention, with special care.

Although the prediction for the next months is optimistic (also supported by the seasonality effect), given the gradual decline of natural and acquired immunity, as well as the high rate of re-infections related to Omicron 2, in order to reduce the risk of extensive flare-up of the pandemic and to return and sustain de-escalation, it is important to:

(a) continue at a high rate both the booster vaccination and the rapid completion of school-age vaccination; and

(b) continue to responsibly comply with the measures currently in force until de-escalation of the dispersion below 5000 per day is observed, including the high number of the self-tests.

Figure 1. Evolution of the COVID-19 dynamics for Greece, as well as the long-term prediction including (a) the number of new infections per day and (b) the daily number of infected in critical condition (intubated in ICU)

References

CHAPIZANIS, D., KARAKITSIOS, S., GOTTI, A. andamp; SARIGIANNIS, D. A. 2021. Assessing personal exposure using Agent Based Modelling informed by sensors technology. *Environmental Research,* 192, 110141.

LYNGSE, F. P., KIRKEBY, C. T., DENWOOD, M., CHRISTIANSEN, L. E., MØLBAK, K., MØLLER, C. H., SKOV, R. L., KRAUSE, T. G., RASMUSSEN, M., SIEBER, R. N., JOHANNESEN, T. B., LILLEBAEK, T., FONAGER, J., FOMSGAARD, A., MØLLER, F. T., STEGGER, M., OVERVAD, M., SPIESS, K. andamp; MORTENSEN, L. H. 2022. Transmission of SARS-CoV-2 Omicron VOC subvariants BA.1 and BA.2: Evidence from Danish Households. *medRxiv*, 2022.01.28.22270044.

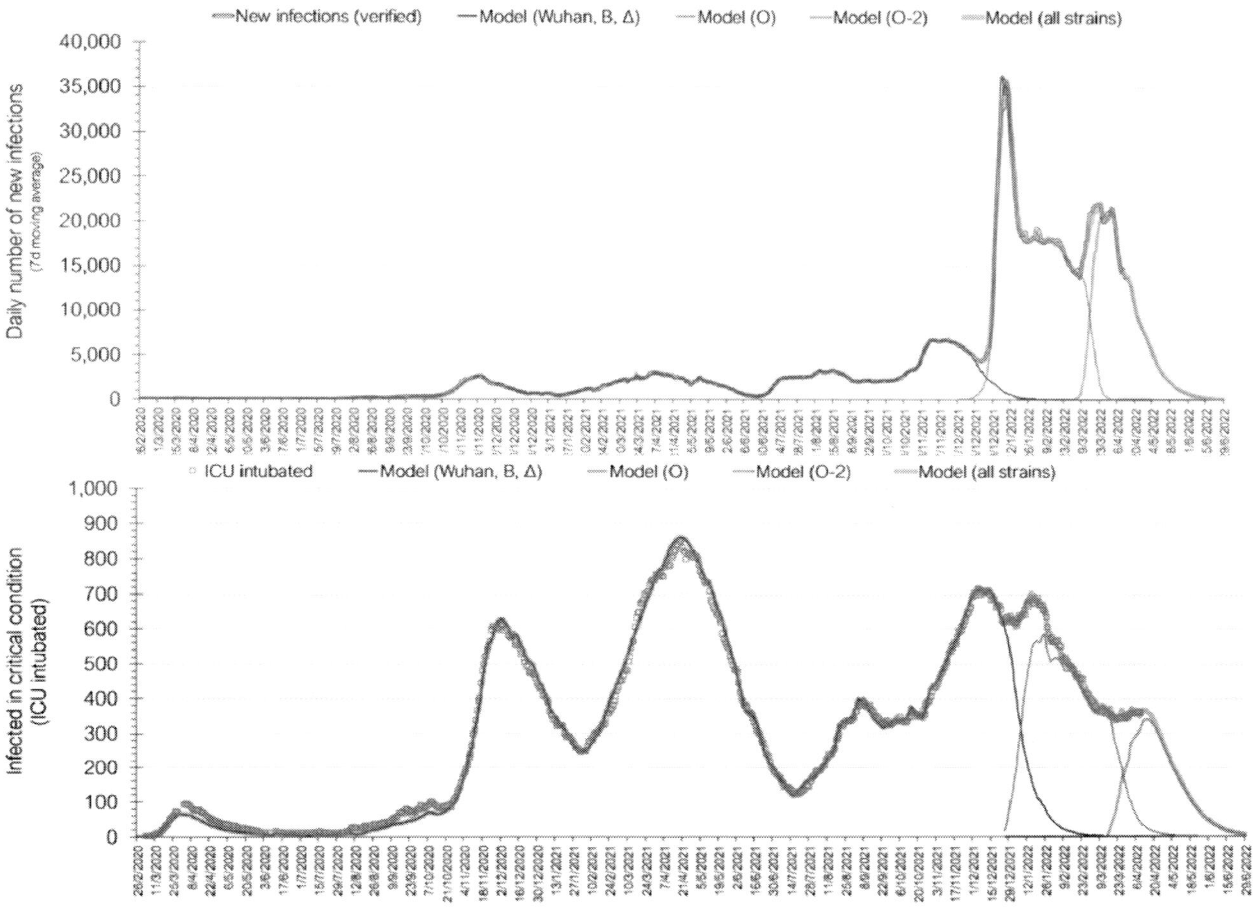
New infections (verified)
Model (Wuhan, B, Δ)
Model (O)
Model (O-2)
Model (all strains)
Daily number of new infections
(7d moving average)
40,000
35,000
30,000
25,000
20,000
15,000
10,000
5,000
0
ICU intubated
Model (Wuhan, B, Δ)
Model (O)
Model (O-2)
Model (all strains)
Infected in critical condition
(ICU intubated)
1,000
900
800
700
600
500
400
300
200
100
0

348c Portable and Label-Free Quantitative Loop-Mediated Isothermal Amplification (LF-qLamp) for Reliable COVID-19 Diagnostics in Three Minutes of Reaction Time

Sergio Bravo-González, Grissel Trujillo de Santiago, Mario Moisés Álvarez

Loop-mediated isothermal amplification (LAMP), and particularly colorimetric LAMP, has been recently studied as an alternative method for cost-effective diagnostics in the context of the current COVID-19 pandemic. Recent reports document that LAMP-based colorimetric diagnostic methods have a comparable sensitivity and specificity to that of RT-qPCR. We report the use of a portable Arduino-based and label-free quantitative LAMP-based amplification system (qLAMP) assisted by pH microelectrodes for the accurate and reliable diagnosis of SARS-CoV-2 during the first 3 min of the amplification reaction.

The discrimination between negative samples and positive samples with medium to high viral loads (i.e., at least 100 copies) was feasible by direct comparison of the pH (i.e., the electric potential trajectories (ΔP)). However, samples containing a low copy number (i.e., less than 10 gene copies) cannot be distinguished from negative samples by only comparing trajectories (Figure 1). The behavior of the trajectory of the amplification reaction in positive and negative samples can be better discriminated during the first stage of the amplification (i.e., the first three minutes of the amplification). Indeed, the value of the area under the curve or of ΔP versus t is distinguishably higher in positive samples than in negative samples during the first stage of the amplification.

We show that this simple system and analysis strategy enables a straightforward discrimination between samples containing or not containing artificial SARS-CoV-2 genetic material in the range of 10 to 10,000 copies per 50 μL of reaction mix. We also spiked saliva samples with SARS-CoV-2 synthetic material and corroborated that the LAMP reaction can be successfully monitored in real time using microelectrodes in saliva samples as well.

In addition, we analyzed a limited set of saliva samples (n=15) from volunteers displaying COVID-19 symptoms. We complemented this experimental set with saliva samples from 30 healthy volunteers. From these experiments with actual saliva samples, we determined a sensitivity >95% and a selectivity >92% for qLAMP.

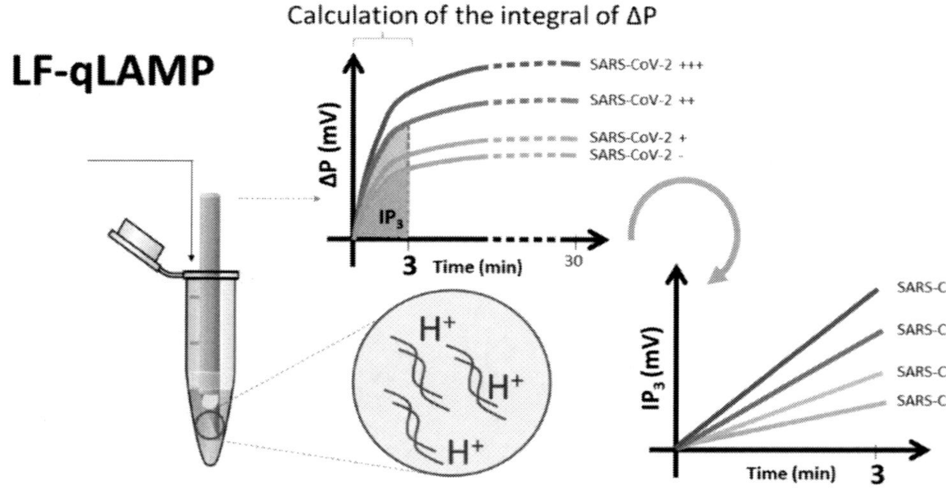

Figure 1. A quantitative LAMP-based system (qLAMP) to identify SARS-CoV-2 genetic material within 3 minutes.

348d Affordable Rapid PCR-Based Biodetection at the Point of Care Via Rayleigh-Bénard Convection

MinGin Kim, Victor Ugaz

Nucleic acid testing approaches like the PCR and its variants are analytical gold standards that play a critical role in diagnosing and monitoring infectious diseases. But PCR-based testing is primarily performed in laboratory settings because the required thermocycling instrumentation is neither amenable to nor validated for operation in a portable format, limiting rapid detection in remote locations and low-middle income countries. A lack of quantitative data to support validation as accepted laboratory methods is a barrier to adopting portable systems for routine use at the point of care (POC). Here we describe a microfluidic POC PCR platform that overcomes this barrier by exploiting the ability to isothermally perform PCR via natural convection [1-2]. This platform can be easily and cost-effectively manufactured without complex equipment and processes, dramatically increasing simplicity, portability, and affordability. The condition to achieve 100% repeatability is identified, providing critical quantitative data to support the adoption of portable convective thermocycling platforms as a centerpiece of validated laboratory methods. We evaluate the platform's performance by determining the detection limit, the minimum reaction time required, and the potential for multiplex analysis. We apply this platform to demonstrate rapid replication of targets associated with multiple pathogens, including SARS-CoV-2. These results are comparable to or surpass those obtained using the conventional benchtop PCR thermocyclers, supporting the high potential of this platform to enable affordable POC nucleic acid-based analysis.

[1] R. Muddu et al., "Chaotically accelerated polymerase chain reaction by microscale Rayleigh-Bénard convection." Angewandte Chemie, International ed.) Vol. 50 (2011): 3048-52.

[2] A. Priye et al., "Microscale chaotic advection enables robust convective DNA replication." Analytical Chemistry Vol. 85 (2013) 10536-41.

348f Rapid Bioproduction of Protein Biologics Using Plants in Response to Sars-Cov-2

Katherine Haddad, Seongwon Jung, Imran Khan, Nancy E. Lane, Somen Nandi, Karen A. McDonald

Problem: As demonstrated by the recent pandemic, the rapid development of scalable protein therapies is critical. At times or places when highly specialized and large-scale equipment is unavailable, e.g. third world countries or space, this need to produce therapeutics and diagnostics is even more imperative. Fortunately, plants are capable of growth in low resource environments for cost-effective and scalable protein production. Plants can be harnessed for biological growth in the same capacity as a bioreactor-based system, however, the current gold standard for recombinant protein production in plants uses gram negative *Agrobacterium tumefaciens* (*A. tumefaciens*) and is optimized for expression in *Nicotiana benthamiana* (*N. benthamiana*), a relative of tobacco, making biologics produced in this system more difficult to quickly and safely purify for clinical use. In contrast, edible plants that are generally regarded as safe (GRAS) can be used as a host for numerous biologics, like lettuce (*Lactuca sativa*), which has been extensively studied for space applications and grown on the International Space Station because it can utilize limited in situ resources for growth. Current production systems require multiple days to produce recombinant proteins, thus reducing the ability to keep up with the rapidly changing needs of global health, as well as the amount of therapeutic which can be brought to market quickly. This project focuses on developing bioprocessing technologies for a dynamic plant model using a viral expression system that can be employed to rapidly produce protein biologics and reagents in less than 24 hours under low resource requirements.

Methods: Romaine lettuce (*Lactuca sativa* var. *longifolia*) and control *N. benthamiana* were transfected by particle bombardment; recombinant protein production dynamics were compared to transformation by infiltration of *A. tumefaciens*. Production of target proteins in leaf tissue extract was identified by visualization on SDS-PAGE and Western blot. The transgene for the protein of interest, the receptor-binding domain protein (RBD) of SARS-CoV-2, was labeled with a His-tag and inserted into a geminiviral vector for rapid production in the plant hosts and used in both particle bombardment and agroinfiltration methods for transient expression.

Results: From the studies, it was identified that bombarded tissue displays variations in protein production depending on age of the plant or growth of the leaf. *A. tumefaciens* infiltration protein dynamics indicate that the genetic construct can efficiently be expressed in both hosts within 24 hours, with *N. benthamiana* expressing the protein of interest in as little as 4 hours (**Fig. 1**).

Implications: While the production of RBD is the focus of this study, the platform is designed to be applicable to produce other protein-based therapeutics and diagnostics within 24 hours. RBD is a valuable initial target as it can be used for screening antibodies and additional therapeutics for diagnostics to combat COVID-19, thus the low-cost and large-scale production of this biologic is imperative. The next step for this project would be to construct a viral vector for systemic infection within the host plants for rapid and enhanced bioproduction of target proteins. This technology has the potential to allow for a mobile therapeutic production system that can be rapidly deployed with limited need for external resources, decreased specialized training for processing, and a large reduction in waste compared to current production platforms.

Figure 1. Time study of agroinfiltrated progression of RBD production in *N. benthamiana* (A) and romaine (B). Lanes 1: protein standard ladder; 2: 4-hours post infiltration (HPI); 3: 8 HPI; 4: 12 HPI; 5: 16 HPI; 6: 20 HPI; 7: wild type plant crude extract; 8: RBD-His positive control, 30ng. Western blot probed with mouse anti-His antibody at a concentration of [1ng/uL] followed by anti-mouse-HRP at [0.1ng/uL].

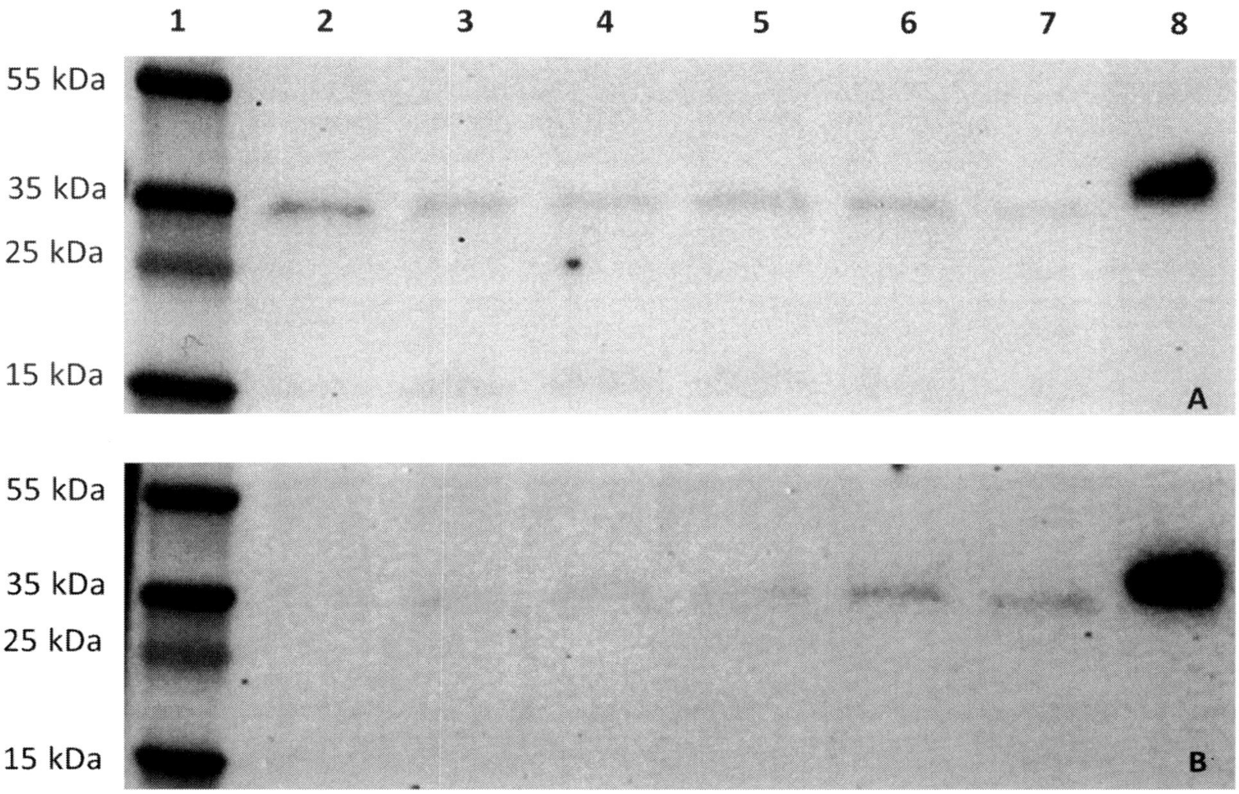
1
2
3
4
5
6
7
8
55 kDa
35 kDa
25 kDa
15 kDa
A
55 kDa
35 kDa
25 kDa
15 kDa
B

348g Strategic Planning of Joint COVID-19 Booster and Influenza Vaccination Campaign: The UK COVID-19 Pandemic Study

Dauda Ibrahim, Zoltán Kis, Maria Papathanasiou, Cleo Kontoravdi, Benoit Chachuat, Nilay Shah, Miriam Sarkis

The emergence of Severe Acute Respiratory Syndrome Corona Virus 2 (aka corona virus disease 2019 or COVID-19) in Wuhan, China and its rapid spread across the world have led the World Health Organisation (WHO) to declare COVID-19 a pandemic on 11 March 2020. To date, globally, COVID-19 have claimed the lives of around 6.2 million individuals and around 58.1 million individuals are still infected with the highly contagious virus, raising concerns on public health. Supported by governments and non-governmental organisations, pharmaceutical companies and academic institutions have deployed a range of platform technologies to produce COVID-19 vaccines in record time in order to mitigate the impact of the virus. The vaccines developed include: (i) viral vectors such as AZD1222 (ChAdOx1 nCoV2) developed by University of Oxford and AstraZeneca, Ad26.COV2.S developed by Johnson and Johnson, Sputnik V (rAd26 and rAd5) developed by Gamaleya Research Institute; (ii) nucleic acid-based vaccines such as BNT162b2 developed by Pfizer and BioNTech, mRNA-1273 developed by Moderna, CVnCoV developed by CureVac; (iii) inactivated pathogens such as BBIBP-CorV developed by Sinopharm; and (iv) protein subunits such as NVX-CoV2373 developed by Novavax.

Despite the successes recorded in vaccination of individuals against COVID-19 (especially in high-income countries), the disease is still prevalent around the globe. This is partly due to the emergence of new variants that render the current vaccines less effective. Booster jabs have been developed and administered to high-risk and vulnerable individuals to avoid developing complications that could lead to hospitalisation or even death. Similarly, influenza virus has been around for some time, and high-risk and vulnerable individuals are required to take annual flu jab to boost immunity and to reduce the impact of the disease within the society. The simultaneous distribution and administration of COVID-19 booster and influenza vaccines poses a significant logistical challenge. For example, COVID-19 and flu vaccines require different type of storage technologies, especially RNA-based vaccines that need to be stored and transported at ultra-low temperature conditions. Consequently, an effective and efficient planning would be required for the vaccination campaigns to be successful.

This work develops a novel multi-product MILP vaccine supply chain model that can be used to plan and optimise the distribution and administration of both COVID-19 booster and influenza vaccines. In addition to planning of vaccine administration, the supply chain model optimises the transport of vaccines from manufacturing plants to administration points. The structure of the supply chain considered here includes manufacturing plants or import locations, fill-finish plants, central stores, regional stores, and administration points. A manufacturing plant, aka primary manufacturing, consists of several unit operations used in the production of drug substance, which is the main ingredient in vaccines, while a fill-finish plant, also known as secondary manufacturing, inserts formulated vaccines (drug substances and excipients) into sterile glass vials or bags that are further packaged into cartons. The multi-product MILP model minimises total cost incurred over the entire vaccination period, while setting targets for total vaccine doses needed, storage capacity required at central store, regional stores and administration points, and identifies cost-effective transportation routes.

The capability of the proposed model is demonstrated using a real-world case study concerned with COVID-19 booster and influenza vaccination campaigns in the UK. Outcomes from this work can be used by government and policy-makers to (i) set out an effective vaccination strategy focusing on the most vulnerable individuals, (ii) assess workforce requirements and financial planning, (iii) monitor the progress of a vaccination and identify resources needed throughout the vaccination period, including storage equipment (fridge, freezer, and ultra-low temperature freezer), transport modes (plane, trucks, and van), vaccine doses (COVID-19 and influenza vaccines), thermal shippers and dry ice, and healthcare personnel.

573a Low Doses of Paclitaxel Induce a Dormant State in Brain Metastatic Breast Cancer Spheroids

Raghu Vamsi Kondapaneni, Rachel Warren, Shreyas Rao

Surviving residual cells after primary breast cancer treatment are known to contribute to tumor recurrence at secondary organs. These recurrent (metastatic) tumors are highly lethal, as they contribute to 90% of breast cancer related mortalities. Residual cells upon disseminating into metastatic sites can exhibit a dormant state and evade cell death. The mechanisms involved in attaining a dormant state are able to protect cells from immune surveillance and even aid them in gaining resistance to chemotherapeutic drugs, thereby enabling them to survive in a hostile microenvironment. The treatment options available for metastatic breast cancer are very few when compared to the primary tumor. In addition, the mechanisms through which cancer cells survive existing therapies at the secondary organ by exhibiting dormancy is not well understood.

In this study, we demonstrate that brain metastatic breast cancer (BMBC) spheroids are able to survive low doses of chemotherapy drug paclitaxel by exhibiting a dormant state through upregulation of phosphorylated p38 (p-p38). 10,000 MDA-MB-231Br BMBC cells were utilized to prepare spheroids. Spheroids were cultured in different drug concentrations (ranging from 0 to 100 μM) for a period of 7 days. Our results demonstrated that the cross-sectional area of the spheroids cultured in 80 nM paclitaxel remained constant throughout the culture time. We observed a decrease in spheroid areas for drug concentrations > 80 nM. 80 nM paclitaxel treated spheroids contained lower levels of proliferating cells than the control spheroids as demonstrated via Ki67 and EdU staining. The percentage of apoptotic cells increased with increase in drug dosages. We investigated the ratio of % extracellular signal - regulated kinase 1/2 (p-ERK) to % p-p38 positive cells as low ratio of p-ERK to p-p38 has been reported to be a characteristic of the dormant state. The ratio of % p-ERK: % p-p38 positive cells were higher in untreated spheroids compared to 80 nM paclitaxel treated spheroids. In addition, 80 nM paclitaxel treated spheroids attained growth upon the withdrawal of drug demonstrating reversibility of the dormant state. Overall, these results providing insight into regulation of the dormant state mediated via low dose chemotherapy.

573b Engineer CAR-Neutrophils for Targeted Cancer Immunotherapy

Xiaoping Bao

The major focus of our research is to engineer human pluripotent stem cells (hPSCs) for targeted cell and cancer therapies. In this talk, I will focus on our recent development of a simplified, chemically-defined platform for producing definitive hematopoietic stem and progenitor cells (HSPCs) and neutrophils for targeted cancer immunotherapy. Neutrophils, the most abundant white blood cells in the circulation, are closely related to cancer development and progression. Primary neutrophils from healthy donors present potent cytotoxicity against different human cancer cell lines through direct contact and via the generation of reactive oxygen species (ROS). However, due to their short half-life and resistance to genetic modification, neutrophils have not yet been engineered with chimeric antigen receptors (CARs) to enhance their anti-tumor cytotoxicity for targeted immunotherapy. Here, we genetically engineered human pluripotent stem cells with different synthetic CARs and successfully differentiated them into functional neutrophils by implementing an innovative chemically-defined differentiation platform. The resulting CAR-neutrophils presented superior and specific cytotoxicity against tumor cells both in vitro and in vivo. Collectively, we established a new platform for the production of CAR-neutrophils, paving the way to myeloid cell-based therapeutic strategies that would complement and boost current cancer treatment approaches. By reducing the complexity of hematopoietic induction and constructing stable CAR-expressing hPSCs, our approach will offer enhanced understanding of human embryonic hematopoiesis and expedite the development of hPSC-derived off-the-shelf hematopoietic and immune cell products for therapeutic purpose.

573c Immunogenic Treatment for Metastatic Breast Cancer Using Targeted Carbon Nanotube Mediated Photothermal Therapy in Combination with Anti-PD-1

Gabriela Faria, Alexis Woodward, Clément G. Karch, Adam Aissanou, Roger Harrison

Introduction:
The lack of targeting treatments for triple negative breast cancer (TNBC) leads to poor prognosis, especially when metastasis has occurred, which spurs on the development of novel treatment strategies.
Although the development of checkpoint inhibition, such as anti-PD-1, has established a breakthrough in cancer treatment, the lack of responsivity to immunotherapy has been correlated with the absence of tumor associated $CD8^+$ T cells, which characterize non-immunogenic tumors. There is evidence that the ablation of solid tumors generates danger signals and tumor antigens that convert it into an immunogenic environment. Thus, the combination of an ablation method to immunotherapy is promising for the generation of a synergistic effect that is capable of not only treating the primary tumor, but also generating a systemic immune response that suppresses metastasis.
We have developed a targeted photosensitizer for photothermal therapy (PTT) for the efficient ablation of solid tumors based on the functionalization of single-walled carbon nanotubes (SWCNTs) to annexin A5 (ANXA5): SWCNT-ANXA5 conjugate. The peak absorbance of SWCNT in the near-infrared (NIR) range efficiently convert light into thermal energy during PTT, aiming to ablate the primary tumor. ANXA5 is a protein that has specificity for cancer due to the strong binding ($K_d \sim 1$ nM) to externalized phosphatidylserine (PS) on the surface of tumor cells and tumor vasculature. PS is a biomarker of various types of cancers, including TNBC, but is absent on the surface of healthy cells.
Here, we study the *in vivo* parameters that better generate an immune activation through SWCNT-ANXA5 mediated PTT: injection mode and tumor surface temperature. Additionally, we combined the optimized therapy with checkpoint inhibition of anti-PD-1, for a systemic immune response against a metastatic tumor model.

Methods:
ANXA5 was produced in *E. coli* and purified by affinity liquid chromatography in an immobilized-nickel column. The high purity protein (>95%, confirmed by SDS-PAGE electrophoresis) is then linked to (6,5) CoMoCAT® SWCNTs with a linker (DSPE-PEG-MAL, short for 1,2-distearoyl-*sn*-glycero-3-phosphoethanolamine-N-[maleimide(polyethylene glycol)]), which interacts hydrophobically with SWCNTs and bonds to ANXA5 through the cysteine reactive maleimide group. EMT6 cells, a TNBC tumor model, were orthotopically injected in BALB c/J female mice. When the tumors were an average of 5 mm in diameter, the mice were treated with a 980 nm NIR laser at 1 W/cm^2. For the first experiment, SWCNT-ANXA5 was injected either intravenously (IV) or intratumorally (IT), and levels of cytokine TNF-α in the serum 1 day after PTT were measured by ELISA. Secondly, using IT injection, the tumors were subjected to PTT, and treatment ceased when the tumor surface reached different peak temperatures (45, 50, 55 and 60°C), which was measured by a thermal camera. Blood serum was collected 1 day and 7 days after irradiation for analysis of proinflammatory cytokines (TNF-α and IL-12). A long-term survival test was performed using IT injection and irradiation until the tumor surface temperature reached 45°C in combination with checkpoint inhibition of anti-PD-1 (three intraperitonial injections of 200 μg/mouse, 1 and 4 days before irradiation, and 4 days after irradiation). The groups in this study were control (untreated), PTT only, anti-PD-1 only, and PTT + anti-PD-1. Lastly, to study primary tumor control, EMT6 tumor

were irradiated with 980 nm NIR laser at 1 W/cm^2 by holding the tumor surface temperature at 45°C ± 1°C for longer period of times (0, 60, 120 and 300 s) by cycles of on/off with the laser. Tumor volumes were recorded for 15 days after irradiation.
Additionally, a biodistribution study of IT injection and another long-term survival study are being performed. Firstly, EMT6 tumor bearing mice were injected intratumorally with the SWCNT-ANXA5 conjugate treated with the PTT until 45°C. Those mice will be euthanized at 1, 2, 3 and 4 weeks after irradiation and major organs harvested for the measurement of SWCNT. Secondly, either EMT6 or 4T1 tumor bearing mice will be treated similarly to the forementioned long-term survival study with the combination of PTT and anti-PD-1, but the PTT will be performed holding the tumor surface temperature at 45°C ± 1°C for 300 s (5 min) by cycling the laser irradiation on and off.

Results:
IT injection was the only method that induced a statistically significant increase in the level of TNF-α in the blood stream ($n = 4\text{-}5$, $p<0.001$). When comparing the irradiation temperatures, 45°C was the group that was more effective in increasing the level of cytokines TNF-α (at 1 day and 7 days after PTT) and IL-12 (1 day after PTT), respectively $p<0.05$, $p<0.01$ and $p<0.01$ for $n=4\text{-}5$. Furthermore, the optimized combination therapy resulted in complete tumor recession and survival at 100 days for 50% of mice in the anti-PD-1 + PTT group. Additionally, the group where the primary tumor was kept at 45°C for 300 s had statistically significant smaller tumors at 15 days compared to the other groups with the tumor held at this temperature for shorter times ($p<0.05$, $n=5$).
The ongoing biodistribution experiment aim to explore whether the SWCNTs remain in the tumor region after intratumoral injection. The long-term survival study intend to assess wheter the irradiation at 45°C for 5 min in combination with anti-PD-1 is efficient at increasing the survival rate of EMT6 tumor model, additionally comparing with the 4T1 tumor model, which is considered less immunogenic.

Implications:
There is evidence that IT injection and PTT at 45°C are more effective at generating immune responses compared to the higher temperatures tested. The lower temperature is important for the safety of the PTT and reduces the chances of damage to surrounding tissue. Lowering PTT temperature could also improve chances of translating this therapy strategy into clinical trials. Increasing the irradiation time holding the tumor surface at 45°C improves the ablation of deeper layers of the solid tumor, improving tumor control. Because EMT6 tumor models are known to establish metastatic tumors very early in the tumor development, the long-term survival is one evidence that the combination therapy of PTT and anti-PD-1 elicits a systemic antitumoral response that efficiently suppresses metastatic tumors.

573d Immune Cell Homing Materials for Cancer Immunotherapy

Hua Wang

Targeted immunomodulation of dendritic cells (DCs) in vivo will enable manipulation of T cell priming and amplification of anticancer immune responses, but a general strategy has been lacking. Here we show that DCs concentrated by a macroporous hydrogel can be metabolically labeled with azido groups in situ, which allows for their subsequent tracking and targeted modulation over time. Azido-labeled DCs were detected in lymph nodes for weeks, and could covalently capture dibenzocyclooctyne (DBCO)-bearing antigens and adjuvants via efficient Click chemistry for improved antigen-specific $CD8^+$ T cell responses and antitumor efficacy. We also show that azido-labeling of DCs allowed for in vitro and in vivo conjugation of DBCO-modified cytokines, including DBCO-IL-15/IL-15Rα, to improve priming of antigen-specific $CD8^+$ T cells. This DC labeling and targeted modulation technology provides an unprecedented strategy to manipulate DCs and regulate DC-T cell interactions in vivo.

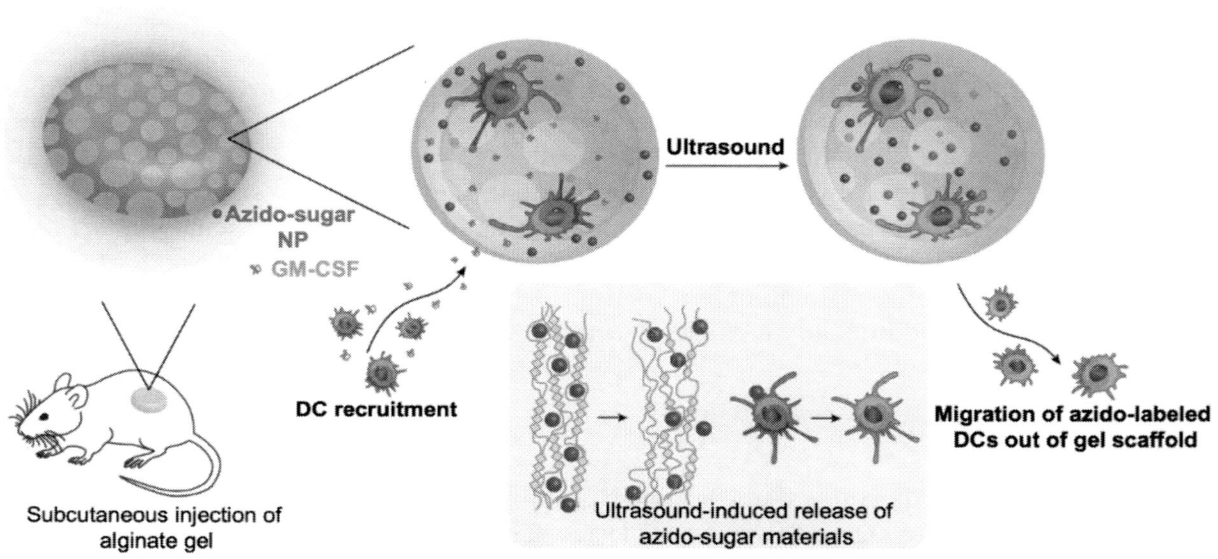

573e Electroporation and Cold Atmospheric Plasma As a Novel Cancer Treatment

Jordan Hoops, Kristen I. Haller, Mikaya M. Elliott, Rylie N. Andrews, Nicole Miller, Timothy Brenza, Prasoon K. Diwakar

Lung cancer is the leading cause of all cancer deaths, with estimates of 130,180 deaths and 236,740 new cases for 2022 in the United States [1]. Available treatments for non-small cell lung cancer include chemo- and immunotherapies which have undesirable side effects such as pain, flu-like symptoms, fatigue, hair loss, anemia, lymphedema, and appetite loss. Current treatments are also limited by toxicity to non-cancerous cells, drug resistance causing immunity to anti-cancer drugs, and length of treatment time. A goal of modern lung cancer treatments is to improve the quality of life given the current side effects. In this study, a novel approach to cancer treatment utilizing membrane electroporation (EP) and administration of cold atmospheric plasma (CAP) is explored. EP delivers electrical pulses to cells, opening pores in the cell membrane while CAP delivers reactive oxygen and nitrogen species at room temperature. The goal of this combination treatment is to take advantage of the disrupted oxidant-antioxidant balance and lowered ROS threshold of cancerous cells, delivering ROS to an apoptotic-inducing state, while causing relatively small damage to healthy cells [2]. In previous studies from Mazandaran University of Medicinal Science, it was observed that CAP administration to B16 melanoma and L929 normal breast tissue cell lines increased cell death in B16 cells and had no significant toxicity in L929 cells [3].

Human epithelial lung cancer cell line, A549, was used in this study. FITC-dextran 4 (FD4) was delivered to cells at varying time intervals with electroporation at various voltages and pulse durations to investigate A549 pore formation and optimal EP parameters. Cellular uptake of FD4 was quantified by fluorescence intensity. Next, cell viability by mitochondrial activity was evaluated following EP treatment and CAP delivery using MTT assay. Future work includes reactive oxygen species quantification, evaluation of cellular progression to apoptosis by caspase 3 activation, and inflammatory cytokine secretion.

[1] Howlader N, et al. National Cancer Institute. 2019.

[2] Krug, D, et al. EEE Transactions on Plasma Science. 2019.

[3] Motaln, H, et al. Molecules. 2021.

573f Combined Gene and Chemotherapy for Cancer: Interdisciplinary Multimodal Approaches to Effective, Safe, and Diverse Therapies

Rebecca Lee, Cheng Wai "Winne" Lei, Margaret Lugin, Jee Young Chung, Woo Chang Hwang, Angela G. Fleischman, Namshik Han, Young Jik Kwon

Most therapies tackle a single pathological target such as an effector protein or a abnormal genetic pathway, often resulting in limited, ancillary therapeutic effects. For example, cancers arise from complex molecular causes, making them difficult to treat by a single modal therapy. In order to rebalance the broken molecular homeostasis in chronic myelogenous leukemia (CML) as a simple, well-documented model, pro-apoptotic and pro-survival genes were simultaneously expressed and silenced in a CML cell, respectively, using a viral/nonviral chimeric nanoparticles (ChNPs) that were composed of the BIM-expressing adeno-associated virus (AAV) core and MCL-1 siRNA-encapsulating, acid-degradable polymeric shell. While BIM/MCL-1 ChNPs genetically rebalance the pathological genetic pathways, a targeted chemotherapeutics for CML, dasatinib, was additionally administered. The result demonstrated synergistically enhanced anti-leukemic effects by the combined multimodal gene and chemotherapy in vitro and in vivo, especially in a rapidly progressing disease stage of blast crisis. A computational simulation via multi-omics network analyses using machine learning also predicted the synergistically augmented apoptosis by the combined multimodal gene and chemotherapy. This study demonstrates the critical roles of engineering therapies that address the complex nature of a disease by integrating interdisciplinary principles.

573g The Landscape of Drug Sensitivity in Cancer Cell Lines Reveals Effective Drug Combinations for Cancer

Belinda Garana, Nicholas Graham

Targeted therapeutics have revolutionized cancer treatment. However, because only seven percent of cancer patients respond to molecularly targeted therapies, improved methods to match tumors with effective therapeutics are desperately needed to realize the promise of precision oncology. In addition, because acquired resistance is almost universally observed in single-agent treatments, it is crucial to identify drug combinations which may increase the efficacy of targeted therapies and/or prevent the emergence of resistant cells. Here, we demonstrate that a bioinformatic algorithm recently developed by our lab, Drug Mechanism Enrichment Analysis (DMEA), can successfully identify effective drug combinations for cancer cells. Leveraging public databases of transcriptomic profiles (e.g., Cancer Cell Line Encyclopedia) and drug sensitivity scores (e.g., Cancer Dependency Map), we analyzed the sensitivity of >300 cancer cell lines to >1,300 drugs to define the global landscape of drug sensitivity. First, we calculated Pearson correlation estimates between the sensitivity scores (AUC) for each drug. Next, we ran DMEA with these Pearson correlation estimates to identify drug mechanisms (e.g., EGFR inhibitors, RAF inhibitors, etc.) enriched based on sensitivity to each drug. Lastly, we ran DMEA again, now ranking the drugs based on the NES scores for each drug mechanism. Finally, the resulting NES scores identified drug mechanisms which may benefit cells sensitive or resistant to each of 85 drug mechanisms. This approach validated that our method can successfully predict sensitivity to ninety nine percent of annotated drug mechanisms. Next, to identify potential drug combinations from our global analysis, we compared the strength and direction of each pair of drug mechanisms. The pairings with the greatest positive scores represented the top potential drug combinations for sensitive cancer cell lines. In our analysis, the top drug combination was inhibitors of polo-like kinase (PLK) and tubulin polymerization. In support of this finding, published studies indicate that tubulin inhibition can increase the toxicity of PLK inhibitors in various cancers. This suggests that other drug combinations predicted by our algorithm may also exhibit synergy against cancer cells. Taken together, our study presents a wealth of information about potential drug combinations for cancer and represents a framework for future analysis of large drug screen datasets.

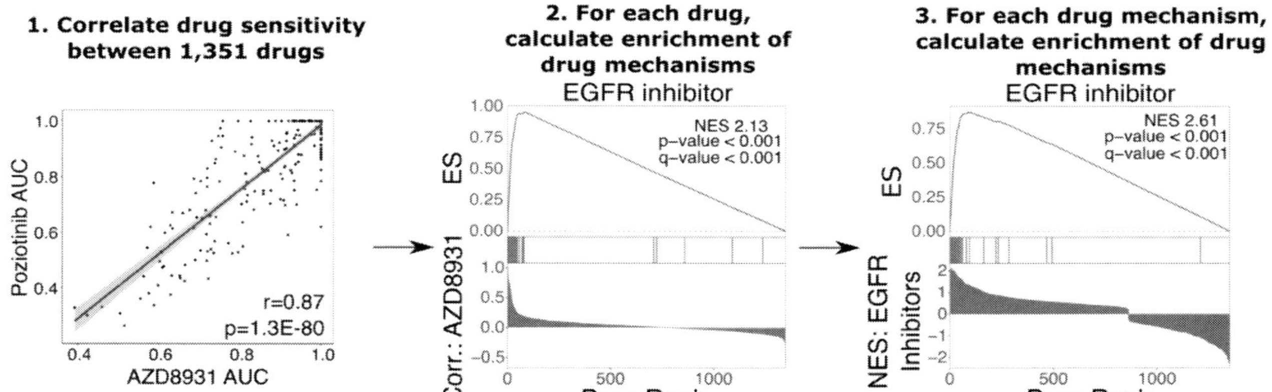

573h Multi-Modal Single-Cell Characterization As a Basis for Precision Therapy for Intratumoral Heterogeneity in Glioblastoma

James Park, Abdullah Feroze, Samuel Emerson, Anca Mihalas, C. Keene, Patrick Cimino, Adrian Lopez Garcia de Lomana, Wei-Ju Wu, Serdar Turkarslan, Kavya Kannan, Nitin Baliga, Anoop Patel

Glioblastoma (GBM) is a heterogeneous tumor made up of cell states that evolve over time. Here, we modeled tumor evolutionary trajectories during standard-of-care treatment using multi-omic single-cell analysis of a primary tumor sample, corresponding mouse xenografts subjected to standard of care therapy, and the matched recurrent tumor collected at autopsy. We analyzed the single-cell, multi-omic data with the Mechanistic Inference of Node-Edge Relationships (MINER) to identify a network of 52 regulators that mediated treatment-induced shifts in xenograft tumor-cell states, several of which were also reflected in recurrence. By integrating MINER-derived regulatory network information with transcription factor accessibility deviations derived from single-cell ATAC-seq data, we developed consensus networks that subtend cell state transitions across subpopulations of primary and recurrent tumor cells. Finally, we identified potential drugs that target regulatory network modules that were active at distinct stages of disease/post-treatment progression. Together, these results support a framework for applying single-cell-based approaches that would enable precision medicine applications to an individual patient in a concurrent, adjuvant, or recurrent setting.

417a Two Complementary Memory B Cell Processes Generate Antibody Protection Against the Same or Variant Viruses

Matthew Vanbeek

We study an important question in immunology: how are memory B cells recalled upon re-exposure to the same or variant antigens? B cells and their antibody products help protect humans from diverse pathogens. Upon infection or vaccination, B cells undergo a Darwinian evolutionary process in germinal centers (GCs) called affinity maturation. During this process, the B cell receptor (later secreted as antibodies) undergoes rapid mutation and selection based off the receptor's ability to bind to a pathogen. This process produces memory B cells poised to respond to future pathogens and plasma cells that secrete the B cell receptor as antibodies. We developed a physics-based computational model incorporating 1) the kinetics of antigen capture, 2) selectivity of T cells, 3) differing affinity between B cell receptors and variants and 4) other relevant biological features to study how memory B cells are recalled upon re-infection or a booster vaccination. We find that prime and secondary GCs generate a diverse pool of B cells, some with low antigen affinity and some with high affinity. Upon re-exposure to the same or similar antigen, memory B cells that enter secondary GCs can undergo mutation and selection to generate even higher affinity B cells, albeit slowly. Meanwhile, upon re-exposure to the same antigen, affinity-dependent selective expansion of available memory B cells outside GCs (EGCs) results in a rapid antibody response comprised of the best available B cells. The EGC response behaves similarly toward a variant antigen, selecting cross-reactive B cells from the diverse pool generated in the prime GC to provide the best possible antibody response while new secondary GCs slowly generate a tailored response for the new variant. The EGC also generates high antibody titers that cause the GC to increase B cell diversity. Our results were validated with data from humans vaccinated with two doses of the same antigen (COVID-19 vaccine). Based on a simple evolutionary model, we suggest that the complementary roles of EGC and GC processes evolved in response to exposure to evolving pathogen families.

417b A Model for How T Cell-Mediated Autoimmunity Can be Triggered By Persistent Viral Infections

Rose Yin, Samuel Melton, Eric Huseby, Mehran Kardar, Arup K. Chakraborty

It has long been known that certain persistent infections can trigger the onset of T cell-mediated autoimmune diseases, but the reasons underlying this phenomenon remain unclear. T cell development in the thymus contributes to creating a largely self-tolerant and pathogen-specific mature repertoire. Some autoreactive T cells survive thymic development and circulate in peripheral tissues, but they usually do not result in autoimmunity. Theoretical and experimental studies suggest that activation of autoreactive T cells does not necessarily lead to a full-blown autoimmune response because a threshold number of T cells need to be activated in response to antigen for T cells to proliferate and differentiate into effector cells. Activation of such a "quorum" of T cells is more likely for foreign antigens compared to self-antigens because of the bias against autoreactive T cells conferred during thymic development. We developed a model for thymic development and then challenged the resulting mature T cell repertoire with varying intensities of infection. Persistent or intense infections were modeled by presenting increasing numbers of foreign antigens. Our results describe key parameters and conditions that result in persistent infections triggering T cell-mediated autoimmunity. Specifically, we describe how T cell activation by multiple foreign antigens can result in weakly autoreactive T cells exceeding the quorum threshold and mounting a response to self-antigens. These results highlight the importance of collective effects for T cell-mediated immunity and its aberrant regulation. Implications of our findings for phenomena such as epitope spreading, as well as possible experimental tests, will also be discussed.

417c Predicting the Effects of Sars-Cov-2 Spike Protein Mutations to MHC Class II – Mediated Immune Responses

Mercedes Haley, Sumaiya Islam, Robert Pantazes

Starting in 2019, the Coronavirus Disease (COVID-19) pandemic spread rapidly, causing wide-spread disruptions across the world. Over time, numerous variants of concern (VOC) emerged as the causative agent of the pandemic, Severe Acute Respiratory Syndrome Coronavirus 2 (SARS-CoV-2), mutated. These mutations all occurred because they provided evolutionary advantages to the virus, but the specific advantages varied. Reasons for mutation include, but are not limited to, increasing infectivity, enhancing survival in the environment, and evasion of immune responses. The focus of this talk is on understanding how SARS-CoV-2 mutations affected immune responses in different ethnic populations.

The adaptive immune response is mediated by the Major Histocompatibility Complex (MHC) proteins, which are coded for by Human Leukocyte Antigen (HLA) genes. The SARS-CoV-2 mutations affect and can disrupt the HLA-driven immune recognition. To understand these effects, we used ProPred to predict the change in binding for every point mutation on the Spike (S) protein of SARS-CoV-2 with respect to the wildtype, providing insight on how mutations increase or decrease viral immunogenicity. Further, HLA alleles appear in different frequencies among ethnic groups. Data from Allele Frequency Net was used to predict how the immune-response altering mutations impacted different ethnic populations.

The most commonly discussed SARS-CoV-2 VOC are the Alpha, Delta, and Omicron strains. Our data show that Delta had more immune response reducing mutations than the other two VOC, which is likely associated with its higher viral load. Alpha, in turn, had mutations that had less impact on the immune response but are known to provide other evolutionary benefits (e.g. increased binding to cellular receptors). Finally, the high number of Omicron mutations helped it evade existing immune responses, from either vaccines or prior infections, but actually increased its overall immunogenicity. This presentation will describe our methods for predicting the immunogenicity impacts of SARS-CoV-2 mutations, provide an overview of the mutations in general, and describe the detailed impacts of their effects on different ethnic populations.

417d Inhibition of Sars-Cov-2 Spike Protein Pseudotyped Virus Infection Using ACE2-Tethered Micro/Nanoparticles

Soha Alkhaldi, Ian Peng, Esmael Alyami, Ammar Ahmad Tarar, Ching-An Peng

Coronavirus disease 2019 (COVID-19) has caused a global pandemic of severe acute respiratory syndrome coronavirus 2 (SARS-CoV-2). Angiotensin-converting enzyme 2 receptor (ACE2) is a key receptor for SARS-CoV-2 to infect the host. The binding between spike protein (S) and ACE2 facilitates the entrance of SARS-CoV-2 to the host cell. SARS-CoV-2 is constantly nutating into new variants that cause high infection rate. The development of prophylactic and therapeutics approaches remains needed to fight against mutated SARS-CoV-2 variants. In this study, ACE2-streptavidin fusion protein expressed by recombinant DNA technology was anchored on biotinylated fluorescent polystyrene particles of various sizes ranging from 0.15 to 5 μm. ACE2-tethered micro/nanoparticles were shown to block spike protein pseudotyped lentivirus entry to ACE2-expressing HEK293T cells. interfere with viral attachment, viral entry, and ensuing infection. Compared to ACE2 in soluble form, 5-μm particles immobilized with ACE2 were more efficiently interfere with viral attachment, entry, and its ensuing infection. Furthermore, ACE2-tethered particles inhibited the infection of SARS-CoV-2 D614G variant up to 90%. Our results showed that particles functionalized with ACE2 could be used as efficient decoys to block the infection of SARS-CoV-2 strains.

417e A Novel SERS-Based Assay for Detection of the Sars-Cov-2 Spike Protein

Karen Wawrousek, Moein Mohammadi, Delphine Antoine, Madison Vitt, Sharmin Sultana Jyoti, Julia Dickie, Gerard Wall, Patrick A. Johnson

The COVID-19 pandemic has highlighted the need for the rapid development of sensitive diagnostic tests. We have developed a rapid, point-of-care assay to detect the spike protein of the SARS-CoV-2 virus. The assay relies on the use of scFv (single chain fragment variable) antibody fragments that bind the receptor binding domain of the SARS-CoV-2 spike protein, with immunocomplex detection using SERS (surface enhanced Raman spectroscopy). ScFv antibody fragments can be expressed in *E. coli*, making this approach less expensive than antibody production in animals or cell culture, while the sensitive SERS detection makes this assay more sensitive than lateral flow assays currently on the market. The assay has a limit of detection of $2x10^3$ SARS-CoV-2 genomes/ml or 5 fg/ml SARS-CoV-2 spike protein, which is below the viral load of an infectious individual. The assay recognizes the spike proteins of the alpha, beta, gamma, delta, and omicron variants of SARS-CoV-2, but does not recognize the closely related SARS-CoV, MERS-CoV, or common coronavirus strain HKU1 spike proteins. In this talk, I will discuss development and validation of the assay, as well as further development and improvement for our SERS-based SARS-CoV-2 detection.

417f Cell-Free, Dendritic Cell-Mimicking Vaccines for Cancer, COVID-19, and Beyond

Melissa N. Thone, Jee Young Chung, Young Jik Kwon

Vaccines are acknowledged as one of the indispensable means for human health, especially during the COVID-19 pandemic. Traditional vaccines are effective and reliable, but vaccines against many infectious and non-infectious diseases are still in great demand. A key orchestrator in the adaptive immune system is dendritic cells (DCs), antigen presenting cells that prime T cells for directed cellular and humoral immunity. DC-based vaccines avoid many hurdles of administration, circulation, uptake by APCs, intracellular antigen processing, and antigen presentation along with stimulatory signals, resulting in orders of magnitude improved vaccination efficacy while lowering side effects. However, DC-based vaccines also offer limitations of physical and biological stability, adaption to suppressive environment, and specific storage and handling requirements. In order to overcome the shortcomings, DCs were chemically converted to acellular vesicles that still preserve most characterizations and functions of DCs, called extracellular blebs (EBs). DC-EBs are structurally and functionally homogenous after produced by fast, simple, and highly scalable processes. They also demonstrated stable antigen presentation to T cells with maturation-dependent tunable stimulation. DC-EBs loaded with a tumor-specific peptide effectively eradicated target tumor cells in vitro and in vivo by generating cytotoxic T lymphocytes (CTLs). EBs derived from DCs genetically engineered to express SARS-CoV-2 spike proteins were able to generate antibodies in vivo that effectively neutralized SARS-CoV-2. EBs are cell-derived materials that allow engineering a broad range of cells to tunable, biologically safe, easy handled, and diverse therapies including vaccines, as an example in this study, and beyond.

417g Agent-Based Modeling of Sars-Cov-2 Transmission during the COVID-19 Pandemic at the University of Delaware

Soham Jariwala, Norman J. Wagner, Antony Beris, Richard Suminski, Gregory Dominick

Decision making and policy setting by universities and localities during the Covid-19 pandemic requires knowledge of how people move and interact in their environment. This program adapts new scientific approaches in agent-based population balance modeling developed under the auspices of the National Science Foundation (RAPID CBET #2040503) to model human movement and interaction at the University of Delaware with the goal of providing new tools to help in developing rational strategies for mitigation and control of the transmission of the novel corona virus, as well as future biological threats. A netlogo model was developed for the University of Delaware campus. Data for the model input are obtained from high-definition video footage of public, outdoor areas including green spaces/parks, sidewalks/streets, and campus walkways/congregating spaces analyzed by artificial intelligence algorithms [1, 2]. The model treats a subset of the students and follows their movement in 15 minute steps throughout the semester with allowances for viral transmission rates, mask wearing and social distancing. Statistical data are collected for a broad range of scenarios and compared against data from UD Covid-19 testing and wastewater testing. The statistical results also enable determining key parameters such as the base replication rate R0 needed for more traditional SEIR type epidemiological models. The results of this agent-based epidemiological modeling and surveillance data collection were developed into a dashboard for use by policy makers as well as for public education about various mitigation strategies. The results of this NSF RAPID project are documented on the web site https://sites.udel.edu/udcovidmodel/ to provide a computational tool and example for use more broadly by communities and in additional and future, challenging public health issues. In addition, a K-12, weblogo based, educational model was developed for classroom use by K-12 educators and students alike, along with study guides and educational materials.

1. Suminski, R.R., et al., *Direct Observation of COVID-19 Prevention Behaviors and Physical Activity in Public Open Spaces.* International Journal of Environmental Research and Public Health, 2022. **19**(3): p. 1335.
2. Suminski, R.R., G.M. Dominick, and N.J. Wagner, *A Direct Observation Video Method for Describing COVID-19 Transmission Factors on a Micro-Geographical Scale: Viral Transmission (VT)-Scan.* International Journal of Environmental Research and Public Health, 2021. **18**(17): p. 9329.

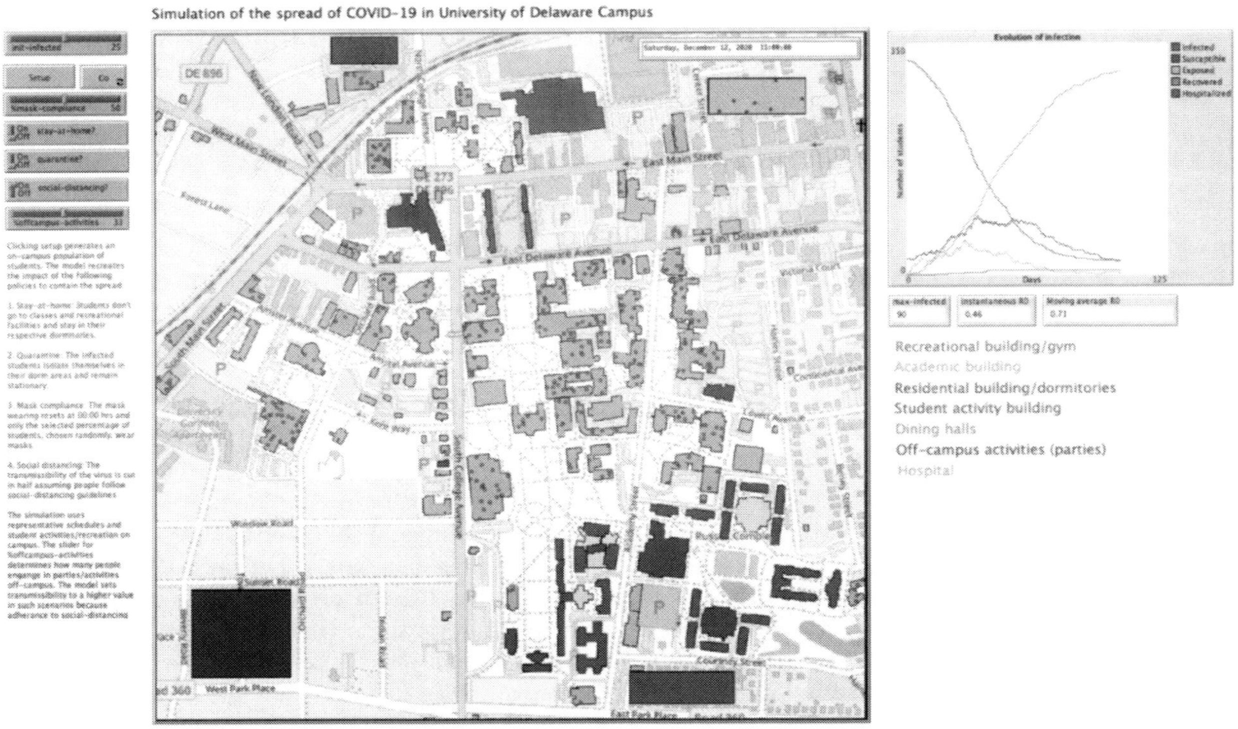
Simulation of the spread of COVID-19 in University of Delaware Campus
Evolution of infection
Days
Recreational building/gym
Academic building
Residential building/dormitories
Student activity building
Dining halls
Off-campus activities (parties)
Hospital

616a Nature-Inspired 3D Scaffolds to Improve *Ex Vivo* T-Cell Culturing Environments for Adoptive Cell Transfer Cancer Immunotherapy

Lucy Todd, Matthew Chin, Marc-Olivier Coppens

There are very few people today who have not been affected by a cancer diagnosis either personally or through a loved one. Cancer can often appear as a growing, undefeatable threat. However, the portfolio of life saving treatments continue to expand with ever increasing success rates. One of the most promising cancer immunotherapy treatments on the market and in research today is the Chimeric Antigen Receptor (CAR) T-cell Adoptive Cell Transfer (ACT). During this treatment, a patient's own immune-cells (T-cells) are extracted, proliferated, and activated ex vivo, genetically engineered to attack the specific tumor cells and then re-infused back into the patient. One of the main challenges with this treatment are the large number of activated T-cells required to proliferate.[1] Currently, ex vivo T-cell culturing environments are often 2D, which neglect to facilitate the 3D mechanical and physical interactions required for efficient T-cell proliferation and activation.[2] By applying the Nature Inspired Solution (NIS) methodology developed at the UCL Centre for Nature-Inspired Engineering, various bio-Inspired 3D T-cell culturing scaffolds have been designed.[3]

Ongoing research addresses two fundamental questions, with the goal to improve the design, efficiency, and scalability of these nature-inspired, 3D T-cell culturing environments. Firstly, how the T-cells migrate, proliferate, and activate on the 3D scaffolds and secondly, why. To answer how, a web-based network of the scaffolds integrates both the geometrical characteristics of the scaffolds (such as porosity, surface area and average Gaussian curvature) and biophysical aspects related to the cells (such as proliferation, and activation rates).[4] This network is inspired by NK fitness landscapes, which are a method of connecting various similarly designed units based off their differing properties (such as a series of scaffolds with varying porosity and surface area) and then visual demonstrating which of those changing properties play the largest role in achieving a specified "fitness" parameter (such as T-cell proliferation or activation rates). Iteratively comparing with experimental results, this web-based network informs future scaffold designs to improve predictable T-cell growth on the scaffolds. To answer why the T-cells migrate, proliferate, and activate on the 3D scaffolds in the specific manner demonstrated, inspiration from the cell memory capabilities demonstrated in animal regeneration (such as worm-splicing regeneration) will be examined. This along with the new and upcoming field of "cell-conscience" informs potential explanations behind the T-cell's "motivation" in their movements within the designed culturing environments. With this knowledge we aim to design improved T-cell culturing environments that can be applied within cancer therapy to make CAR-ACT a more affordable and efficient cancer treatment.

(1) Jin, Z.; Li, X.; Zhang, X.; DeSousa, P.; Xu, T.; Wu, A. Engineering the Fate and Function of Human T-Cells via 3D Bioprinting. Biofabrication 2021, 13 (035016). https://doi.org/10.1088/1758-5090/abd56b.

(2) Jensen, C.; Teng, Y. Is It Time to Start Transitioning From 2D to 3D Cell Culture? Frontiers in Molecular Biosciences 2020, 7, 33. https://doi.org/10.3389/FMOLB.2020.00033/BIBTEX.1

(3) Coppens, M.-O. Nature-Inspired Chemical Engineering for Process Intensification. Annu. Rev. Chem. Biomol. Eng. 2021, 12 (1), 187–215. https://doi.org/10.1146/ANNUREV-CHEMBIOENG-060718-030249.

(4) Chin, M. H. W.; Gentleman, E.; Coppens, M. O.; Day, R. M. Rethinking Cancer Immunotherapy by Embracing and Engineering Complexity. Trends Biotechnol. 2020, 38 (10), 1054–1065. https://doi.org/10.1016/j.tibtech.2020.05.003.

616b Isolation and Identification of Subpopulations of Circulating Tumor Cells Using a Microchip with Bio-Inspired Patterns

Wei Li, Zhenya Ding

Circulating tumor cells (CTCs) are highly heterogeneous and contain many cellular subpopulations. It is known that specific CTC subpopulations, rather than the whole, are responsible for cancer metastases. Recently, the phenotype of metastasis-initiating cells (MIC, EpCAM CD44 CD47 MET) among CTCs have been experimentally demonstrated. Emerging microfluidic technologies have shown great promise for the complete capture of the CTCs population with high yield and enhanced purity. However, most existing devices simply isolate all CTCs in a blood sample without resolving them into distinct subpopulations, preventing researchers from acquiring true insights into the metastatic potential CTCs. Isolating, purifying, and profiling individual populations of heterogeneous CTCs remain significant challenges. In addition, cell capture is only the first step in a pipeline, where post-capture, immunofluorescence characterization is required for identification of CTCs from non-specifically adhered blood cells. Since fixation and permeabilization remain standards for CTC characterization, this destructive processing limits the potential for subsequent phenotypic assays and development of personalized disease models for cancer treatments. Quick non-invasive characterization of CTCs and recovery of CTCs with high viability are required for integrating CTC technology to clinical applications. Ideally, platforms for isolation and identification of CTCs need to be incorporated into clinical care settings.

The goal of this work is to develop a new platform for fractionation and profiling of CTC subpopulations and elucidate the metastatic potential of CTCs. To do so, I propose to isolate, in-situ identify, and selectively recover CTCs using a microchip with hyperuniform structure created by microposts (Figure 1). Hyperuniformity (HU) is an emerging concept of a packing pattern which contains local heterogeneity or randomness and global regularity or homogeneity. The HU concept will be integrated for the first time into affinity-based microfluidic devices for CTC isolation. I hypothesize that due to the controlled differences in local flow patterns induced by the hyperuniform structure, cell arrest in different locations on the microchip will require different adhesive strengths. Further, this adhesive strength is anticipated to be related to the types and densities of surface markers on the captured CTCs and, therefore, their metastatic character. Combined with methods to effectively reduce blood cell adhesion and non-invasive release of CTCs, this device can recover individual groups of CTCs on demand for downstream bio-analyses. The HU microchip offers a simple and unique resolution for fractionation of CTCs as its 1) global homogeneity provides equal possibility for all CTCs to adhere; and 2) local heterogeneity allows simultaneous differentiation of subpopulations by analyzing adhesive strength required for individual CTCs. As a result, subpopulations of CTCs can be identified using only their capture locations on the HU chip without requiring additional post-capture immunofluorescence characterization. This HU microchip is expected to profile CTC heterogeneity during cancer metastasis with respect to the numbers/fractions of CTCs with specific surface markers, quantitative expression of specific surface markers that relate to tumor progression, and dynamic change of biophysical properties of CTC subpopulations. This approach is expected to provide new insight into addressing the heterogeneity of CTCs, which will permit CTCs to be isolated and identified nondestructively to ensure that collected downstream genomic and proteomic information reflects the cells' true characteristics, as current CTC characterization methods often created artifacts during sample processing.

Figure 1. Schematics for isolate, in-situ identify, and downstream analysis of CTCs using a microchip with hyperuniform structure created by microposts.

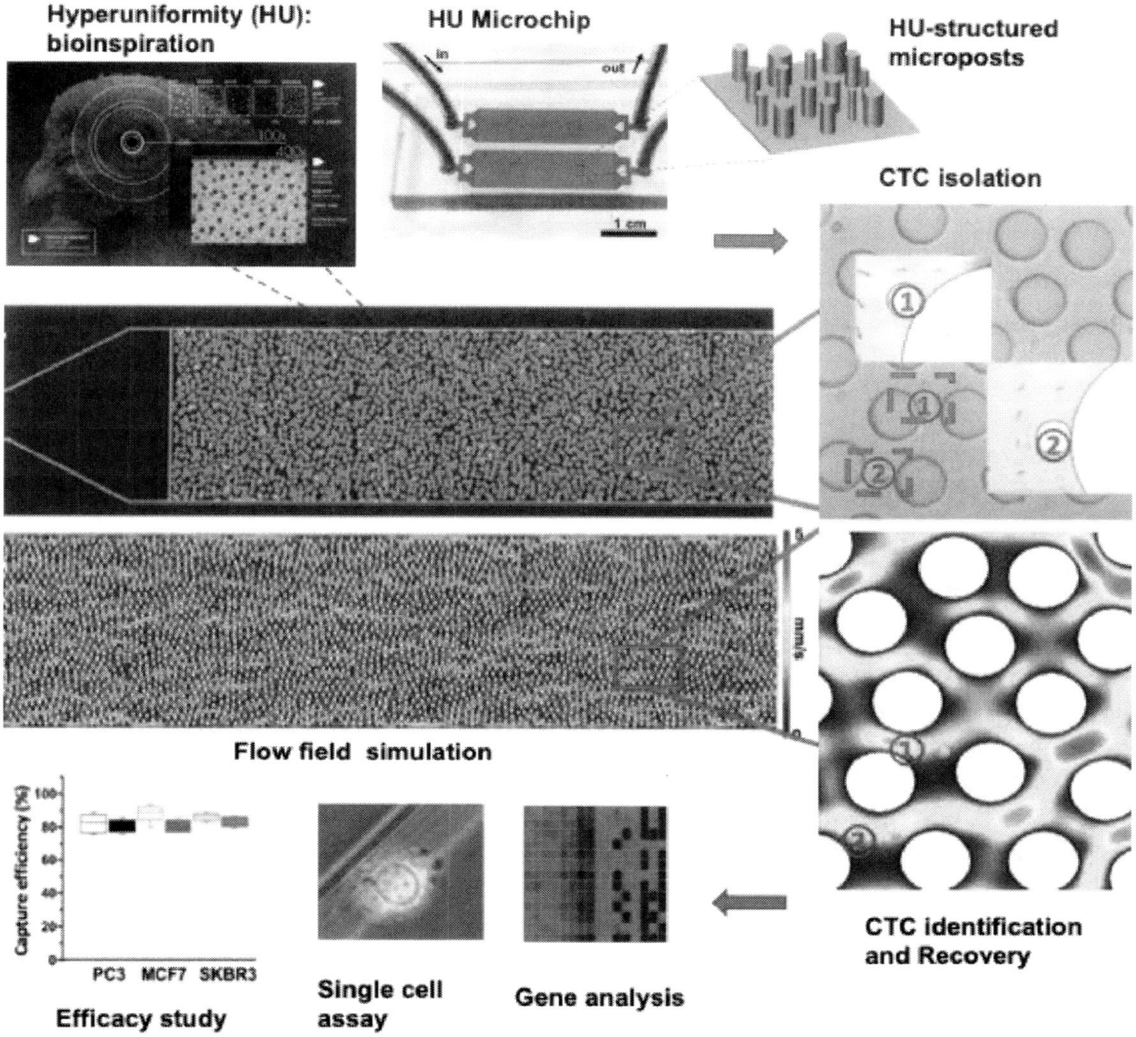
Hyperuniformity (HU): bioinspiration
HU Microchip
in
out
1 cm
HU-structured microposts
CTC isolation
Flow field simulation
CTC identification and Recovery
Capture efficiency (%)
PC3 MCF7 SKBR3
Efficacy study
Single cell assay
Gene analysis

616c Dynamic 3D Co-Culture of Stromal and Immune Cells with ER^+ Breast Cancer Using a Thiol-Acrylate Hydrogel Scaffold and Microfluidic Droplet Trapping Array

Braulio Ortega Quesada, Anowar H. Khan, Sophia Zhou, Elizabeth C. Martin, John Pojman, Adam Melvin

Culturing cancer cells in a three-dimensional (3D) environment better recapitulates in vivo conditions by mimicking cell-to-cell interactions and mass transfer limitations of metabolites, oxygen, and drugs. Recent drug studies have suggested that a high rate of pre-clinical and clinical failures results from mass transfer limitations associated with drug entry into solid tumors that 2D model systems cannot predict. Droplet microfluidic devices offer a promising alternative to grow 3D spheroids from a small number of cells to reduce intratumor heterogeneity which is lacking in other approaches. Many of the existing techniques for 3D cell culture focus on a single cell type; however, this does not adequately represent the tumor microenvironment (TME) which consists of numerous different cells types including stromal and immune cells. Moreover, recent co-culture studies have identified how cancer cells can manipulate nearby health cells to drive both tumor progression and drug resistance; however, many of the underlying mechanisms are still unclear. Our prior work demonstrated the feasibility of incorporating a thiol-acrylate (TA) hydrogel scaffold into a microfluidic droplet trapping array to generate uniform 3D spheroids of a model estrogen receptor-positive (ER^+) cell line (MCF7). The TA hydrogel rapidly (~35 min) polymerized on-chip to provide an initial scaffold to support spheroid development followed by a time-dependent degradation The microfluidic device is coupled with a gravity-driven media infusion system to continually supply the 3D spheroids with fresh media to avoid growth stagnation or cell death. The goal of this work was to expand upon the utilities of our platform to allow for the dynamic 3D co-culture of stromal and immune cells with the MCF7 cells. A strength of the TA hydrogel is that it is transient and will degrade within ~24 hours allowing for the spatial rearrangement of the two different cell types in the spheroid which better recapitulates what occurs in the TME. Three different ratios (1:1, 2:1, and 4:1) of co-cultured primary fibroblasts and ER^+ breast cancer MCF-7 cells showed that increasing the density of the fibroblasts increased the relative size of the spheroid. Spatial tracking of fluorescently-tagged fibroblasts showed that they localized in the center of the spheroid after ~3 days of culture. Terminal immunostaining for the extracellular matrix (ECM) protein Collagen I showed increased expression in the co-cultured spheroids when compared to the mono-cultured MCF-7 spheroid supporting the theory that cancer/fibroblast crosstalk can lead to remodeling of the TME. Co-culture of MCF-7 cells with a model monocyte cell line (THP-1) resulted in the chemical-free differentiation of the THP-1 cells into macrophages as determined by terminal CD11a staining. Unlike the fibroblasts, the macrophages had no spatial limitations within the spheroid and were even found to shed off of the spheroid onto the bottom of the well. Interestingly, the differentiated macrophages exhibited an M2 (or pro-cancer) phenotype when co-cultured with the MCF-7 cells as determined by terminal CD163 staining. These findings support the idea that 3D cultured ER^+ breast cancer can manipulate healthy cells in the TME to adopt a pro-cancer phenotype.

616d Cell-Free Protein Synthesis Biosensing of Glutamine in Human Serum and Saliva: Towards at-Home Low-Cost Diagnostics for Personalized Cancer Treatment

Brad Bundy, J Porter Hunt, Mehran Soltani

Glutamine is the most abundant amino acid in the human body and is essential for metabolism and immune system function. Facile at-home measurements of Glutamine concentrations in bodily fluids could enable personalized highly-effective treatment of cancer, diabetes, infection, and anorexia. To achieve this objective, here we report the development of a lyophilized cell-free protein synthesis sensing platform for rapid detection of glutamine concentrations in human serum and saliva.[1-3] The approach could have a transformative impact in enabling low-cost point-of-care and at-home diagnostics for the personalized dynamic treatment of cancer and other diseases.

References:

1) Hunt JP, Barnett RJ, Robinson H, Soltani M, Nelson JA, Bundy BC. Rapid sensing of clinically relevant glutamine concentrations in human serum with metabolically engineered E. coli-based cell-free protein synthesis. Journal of biotechnology. 2021 Jan 10;325:389-94.

2) Soltani M, Bundy BC. Streamlining cell-free protein synthesis biosensors for use in human fluids: In situ RNase inhibitor production during extract preparation. Biochemical Engineering Journal. 2022 Jan 1;177:108158.

3) Soltani M, Hunt JP, Bundy BC. Rapid RNase inhibitor production to enable low-cost, on-demand cell-free protein synthesis biosensor use in human body fluids. Biotechnology and Bioengineering. 2021 Oct;118(10):3973-83.

616e Investigating Intracellular Adhesion and Intercellular Signaling in Glioblastoma Models

Rosalyn Hatlen, Padmavathy Rajagopalan

Introduction

Glioblastoma (GBM) is the deadliest and most common form of brain cancer. Despite therapeutic intervention, the median survival time of patients is only 15 months. It has been reported in soft-tissue cancers that cells align along with the biomaterial scaffold prior to collective migration. Investigating the intracellular adhesions in GBM cells may help further the understanding of how these cells migrate, as well as elucidate interactions with the matrix. In addition to cell-cell and cell-matrix physical interactions, cytokine and chemokine secretions including TGF-β have been correlated to GBM tumor malignancy and growth. By combining intracellular adhesions and intercellular signaling pathways, it is possible to obtain a clearer picture of interactions between cells in GBM.

Methods

We assembled hydrogels composed of Type 1 collagen or HyStem-C® (Sigma), a gelin and hyaluronic acid (HA)-based biomaterial. A human GBM cell line LN229 (ATCC) was utilized since these cells display a mesenchymal subtype, and are moderately invasive. Layered 3D hydrogels were prepared in glass-bottom tissue culture well plates. First, a thin gel of collagen (either 1.1 mg/mL or 2.2 mg/mL) was deposited to prevent cells from adhering to the glass surface beneath. This was followed by a layer of collagen I or HyStem-C®. LN229 cells were then seeded atop the stacked gels. Co-cultures were also assembled utilizing human umbilical vein endothelial cells (HUVECs) as a mimic of vasculature. The HUVECs were cultured for 7 days to allow the formation of networks before co-culture.

Results

Within 24h in culture, LN229 cells formed bands that were aligned and extended along the z-axis in all biomaterial scaffolds. The adhesion molecule N-cadherin has previously been reported to be expressed by GBM cancer cells in intracellular interactions. Immunostaining revealed LN229 cells expressed N-cadherin within 24 h in culture. The expression of N-cadherin was similar in both mono- and co-cultures as well as across all three biomaterials. In addition to N-cadherin staining, the secretion of TGF-β was also measured, with LN229 monocultures displaying an increase of 9.7 – 22.3-fold in all biomaterials over 7 days of culture. TGF-β secretion in co-cultures increased up to 2.5-fold. TGF-β expression at day 8 was 5.7 – 12.9-fold higher in co-cultures than LN229 monocultures; however, secretion was similar at day 14. Furthermore, increased HUVEC seeding density resulted in a 1.1 – 1.3-fold decrease in TGF-β expression. This suggests that LN229 cells are the primary contributor of TGF-β in co-culture. The expression of chemokine CXCL12 was also measured, with an increase of 2 – 2.6-fold observed from Day 8 to Day 14 in all co-cultures. Increased HUVEC seeding density resulted in up to 1.4-fold increase in CXCL12 secretion. This suggests that HUVECs are in fact the primary source of this chemokine in co-culture. These results indicate that signaling molecules are differentially expressed in co-culture than monoculture.

Ongoing and Future Work

Currently, we are in the process of investigating early and continued expression of N-cadherin. We also plan to examine the expression of additional cell adhesion molecules such as integrins α_5 and β_1, as well as measure the secretion of other cytokines, such as IL10, that could impact intracellular interactions. EC intracellular adhesion will also be investigated through vascular endothelial (VE)-cadherin expression. Investigating how cell adhesion markers and secreted chemokines change as a function of co-culture can lead to a better understanding of intercellular interactions in GBM.

616f Probing the Spatiotemporal Attributes of Gamma-Secretase/Notch Signaling in Breast Cancer Microenvironments

Malcolm Lane Gilchrist, Yueming Li

The overall objective of this research is to investigate the spatiotemporal attributes of the γ-secretase/Notch pathway and inhibition in triple negative breast cancer (TNBC) and provide a molecular basis for therapeutic development. The mortality from TNBC in African American (AA) women is higher than in White American (WA) or European-American (EA) women. Recent studies have indicated that Notch signaling plays an important role in this racial disparity. Notch signaling has been associated with tumor transformation, proliferation, survival, angiogenesis and metastasis. We have shown that γ-secretase cleavage of Notch is highly regulated in breast cancer cells (Villa et al (2014) 8, 1077-1092). Our other recent studies have shown that subtle changes in membrane lipid composition affect the distribution of Notch substrates and γ-secretase and thus enzymatic activity(Barros et al. (2020) Langmuir 36, 23, 6569–6579). At this point, ethnic differences in γ-secretase/Notch signaling and its role in the higher incidence of and poorer outcomes of AA from TNBC are poorly understood.

We hypothesize that γ-secretase plays a major role in the control of Notch signaling in TNBC and contributes to this racial disparity. The emergence of Notch signaling as a therapeutic target of interest in TNBC has spurred interest in this pathway as a potential suspect in the racially disparate burden in TNBC. At this point, ethnic differences in Notch signaling and its role in the higher incidence of and poorer outcomes of African-americans from TNBC remain understudied. In this work we have characterized the Notch and γ-secretase pathways in triple negative breast cancer (TNBC) lipid microenvironments using TNBC-proteoliposomes and TNBC specific lipid formulations in order to examine potential racial disparities at the molecular level in supported bilayer systems. We are comparing these results to our earlier studies of breast cancer cell lines using the same chemical biologic methods and γ-secretase inhibitors (GSIs) (Villa et al (2014) 8, 1077-1092) To further interpret and understand the spatiotemporal distribution and reporter staining patterns of γ-Secretase and the Notch substrates and fragments we have implemented spatial reaction-diffusion computational models. This method is being used to explain the spatial buildup/loading of the relevant signaling-involved molecules and this work will facilitate the examination of racial disparities in Notch and γ-secretase pathways at the molecular level in TNBC and will allow for the examination of how γ-secretase lipid microenvironment is coupled to the mechanism in the context of Notch signaling in TNBC. We believe that this work could lead to the basis for new racially-distinct therapeutic strategies based on γ-secretase inhibition characteristics.

616g Multivariate Model of an Engineered Niche Delineates Metastatic Potential of Breast Cancer

Sophia Orbach, Christian DeVault, Elizabeth Bealer, Jacqueline Jeruss, Lonnie Shea

Background: Approximately 4% of all women in the United States will develop metastatic breast cancer in their lifetime. Currently, metastatic breast cancer has a 10-year survival rate of less than 15%. One of the major inhibitory factors towards improving these outcomes is the inability to detect metastatic disease prior to macroscopic secondary tumor formation, at which point millions of tumor cells have colonized distal sites. Development of metastatic tumors requires the cancer cells and local microenvironment to undergo a complex series of events referred to as the metastatic cascade. Identification of systemic markers correlating with these events could be the key to the early detection, leading to the earlier treatment of metastatic breast cancer and improvements in survival outcomes.

We have designed a porous, polycaprolactone scaffold that acts as a synthetic metastatic niche in mouse models of breast and pancreatic cancer. The scaffolds, when implanted subcutaneously, recruit aggressive populations of tumor cells prior to their detection at native metastatic sites (e.g., lungs and livers). Analysis of the microenvironment at the scaffold can more successfully predict disease progression and recurrence than either the blood or primary tumor. Herein, we employed this scaffold technology to analyze changes in the microenvironment that can delineate metastatic cancer from non-metastatic disease using cell lines derived from a single spontaneous breast tumor in a BALB/cfC3H mouse. Each cell line correlates with a different stage of the metastatic cascade: 67NR is non-metastatic, 4T07 is micro-metastatic, and 4T1 is highly metastatic. By utilizing microenvironmental changes at the scaffold, we developed a 9-gene signature that can effectively identify the metastatic potential of triple-negative breast cancer throughout disease progression.

Methods: Porous, polycaprolactone (PCL) scaffolds were implanted subcutaneously into the backs of female BALB/c mice. After two weeks, mice received an orthotopic inoculation of either the 67NR, 4T07, or 4T1 triple-negative cell line in the fourth, right mammary fat pad. The tumors were allowed to progress for 7, 14, or 21 days before we collected the scaffolds, lungs, primary tumors, and either the spleen or blood (as a systemic control). These time points correlate with the pre-metastatic, micro-metastatic, and metastatic niche in the 4T1 model. At 14 days post-inoculation, tissues were collected for RNA-sequencing analysis using a NovaSeq (S1). The raw counts were normalized by DESeq2 and the normalized tables were used for signaling pathway enrichment (GSEA) and determination of a gene set that isolates the 4T1 samples from the other conditions (mixOmics). Tissues were collected at each time point to investigate immune cell distributions and validate the gene signature with PCR.

Results: We first analyzed the primary tumor to measure its ability to differentiate metastatic potential. Clinically, biopsy of the primary tumor is often the only indicator for invasive cancers if secondary tumors cannot be detected through imaging. The immune cell populations in the 4T07 micro-metastatic and 4T1 metastatic tumors were highly comparable. Of the nine populations tested, only the Gr1+CD11b+ neutrophils were significantly different between the two tumors, which only varied by 28%. In contrast, the 67NR non-metastatic tumors had around 3 times as many F4/80+CD11b+ macrophages, 11-fold less CD11c+F4/80- dendritic cells, 3-fold less Gr1+Cd11b+ neutrophils, and 2-fold less CD4+ T cells. Furthermore, gene expression at the primary tumor was unable to differentiate the 4T07 and 4T1 cell lines. Combined, these data indicate that analysis of the primary tumors cannot identify metastatic potential of invasive cancers.

We next investigated the scaffolds to assess their ability to identify metastatic cancer cell lines. Immune cell populations were unable to distinguish the 4T07 and 4T1 cell lines in either the scaffold or the spleen for the pre- or micro-metastatic conditions, suggesting that immune cell numbers are unable to predict metastasis locally (primary tumor) or systemically (scaffolds and spleen). Interestingly, gene expression at the scaffold was successfully able to predict metastatic potential. Angiogenesis, response to wound healing, monocyte chemotaxis, T cell-mediated cytotoxicity, and epithelial-to-mesenchymal transition were among the cancer-related signaling pathways that were enriched in the 4T1 samples in the scaffold and the lung but were not predicted in the spleen. The gene expression data was then utilized to develop a 9-gene signature that differentiated the 4T1 samples from the healthy (control), 67NR, and 4T07 samples in the scaffold. This signature consisted of *Dhx9*, *Dusp12*, *Fhl1*, *Ifitm1*, *Ndufs1*, *Pja2*, *Slc1a3*, *Soga1*, and *Spon2* and, upon validation with PCR, was able to isolate the 4T1 samples at day 7, day 14, and day 21 (**Figure 1**).

Conclusions: Changes in gene expression at the scaffolds identified the enrichment of cancer-related pathways in response to the inoculation of 4T1 (metastatic) tumors relative to the healthy controls, 67NR (non-metastatic), and 4T07 (micro-metastatic) tumors. Signaling changes at the scaffold mimicked the response at the lung, demonstrating the potential of the scaffold to act as a surrogate tissue for native metastatic sites. We established a 9-gene signature from the scaffold that can identify the metastatic potential of breast cancer prior to the development of macroscopic secondary tumors. Such a signature could classify patients in need of systemic therapy at earlier stages in disease progression, which could improve outcomes.

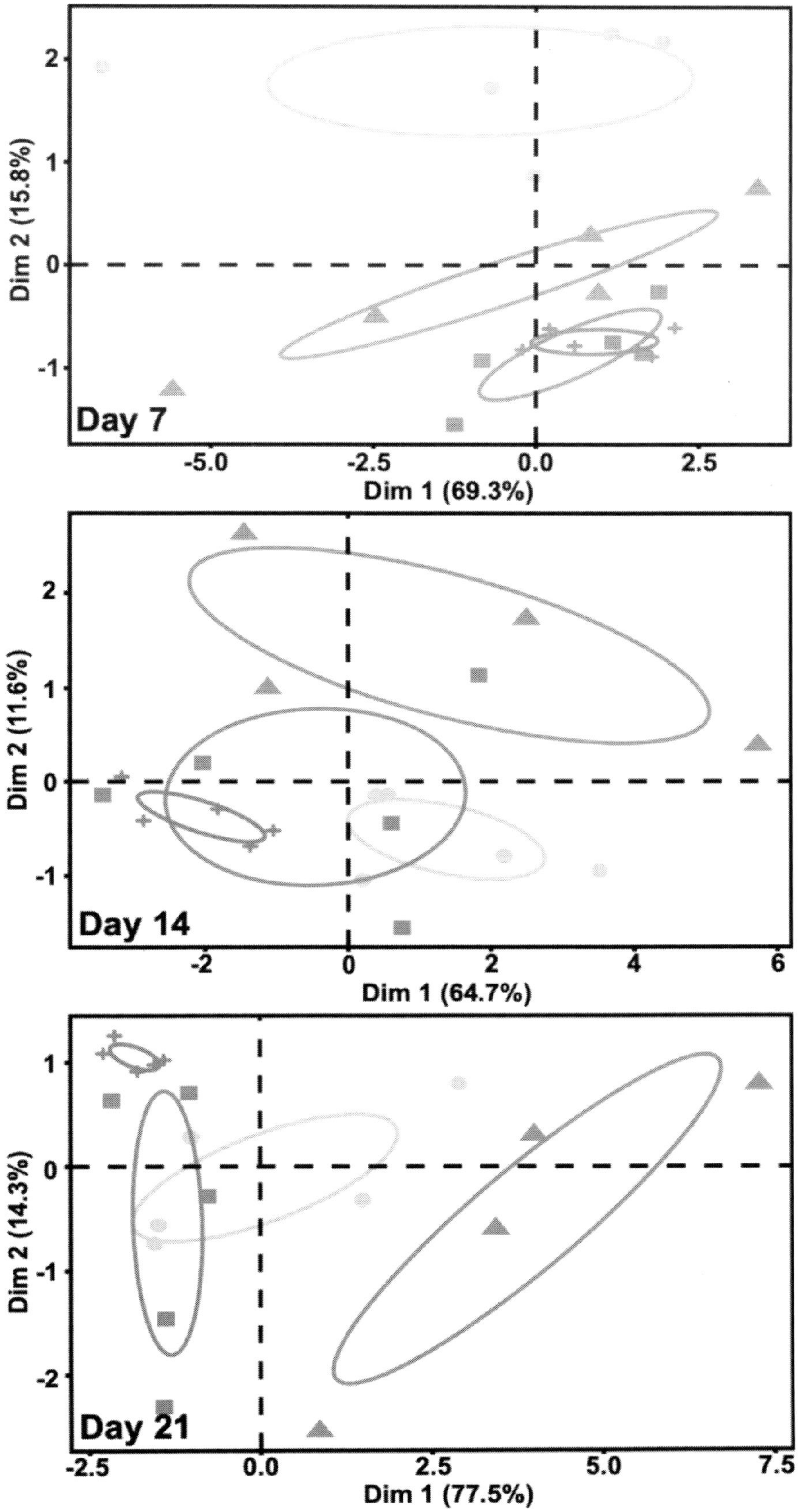

Day 7
Dim 2 (15.8%)
Dim 1 (69.3%)
2
1
0
-1
-5.0
-2.5
0.0
2.5
Day 14
Dim 2 (11.6%)
Dim 1 (64.7%)
2
1
0
-1
-2
0
2
4
6
Day 21
Dim 2 (14.3%)
Dim 1 (77.5%)
1
0
-1
-2
-2.5
0.0
2.5
5.0
7.5

616h Multimodal Imaging for Simultaneous Measurement of Melanocyte Mass and Pigmentation

Tarek Moustafa, Edward Polanco, Rachel Belote, Robert Judson-Torres, Thomas Zangle

Despite advances in treatment, the 5 year mortality rate for patients with advanced melanoma remains at 48%. Potentially effective targeted therapies, e.g. BRAF-inhibitors for patients harboring a BRAF mutation, exist for advanced melanoma. However, a key challenge to reducing melanoma mortality rates is that many patients' tumors will eventually acquire resistance to these therapies. Predicting acquired resistance is, therefore, a major challenge that needs to be overcome to have an impact on patient outcomes in melanoma. We, and others, have previously shown that drug response of individual cells within large populations can be measured using quantitative phase imaging (QPI). QPI is a label-free microscopy method that measures the amount that light slows as it interacts with cell mass. QPI data can be used to measure total cell mass, or the rate of change of cell mass due to growth. Changes in growth rate during treatment with targeted therapies can then be used to monitor cell drug responses, and analyzed for signs of therapeutic resistance. Preliminary data also shows that increases in melanocyte pigmentation is associated with acquired drug resistance. We have, therefore, engineered a system for simultaneous quantification of cell mass, via QPI, and cell pigmentation, via dark field microscopy. In this work, we present the design and validation of this system as a high-throughput screening tool that can be used to study mechanisms of acquired drug resistance in melanoma.

AUTHOR INDEX

Aguirre-Soto, Alan 46
Aissanou, Adam 70
Al'Abri, Ibrahim 27
Alimirzaei, Atefe 35
Alkhaldi, Soha 80
Álvarez, Mario Moisés 62
Alyami, Esmael 80
Andreadis, Stelios 39, 43
Andrechak, Jason C. 32
Andrews, Rylie N. 73
Antaki, James 19
Antoine, Delphine 81
Ayati, Royaalsadat 14
Baliga, Nitin 76
Banerjee, Ipsita 41
Bao, Xiaoping 69
Bealer, Elizabeth 93
Belote, Rachel 96
Beris, Antony 83
Bilodeau, Camille 49
Boucher, Richard C. 4
Bravo-González, Sergio 62
Brennan, Molly 30
Brenza, Timothy 73
Bundy, Brad 89
Bunnett, Nigel 24
Burns, Carolyn A. 58
Carlson, Ross P. 25
Case, Marshall 49
Chachuat, Benoit 66
Chakraborty, Arup K. 78
Chan, Christina 31
Chandler, Kevin 12
Chauhan, Ayushi 15, 28
Chavda, Himani 26
Chen, Kevin 31
Chin, Matthew 85
Choudhury, Debanik 39
Chowdhury, Ratul 48, 55
Chung, Jee Young 74, 82
Cimino, Patrick 76
Coppens, Marc-Olivier 85
Crook, Nathan 27
Danielsen, Scott 4
Dantas, Gautam 1
De Lomana, Adrian Lopez Garcia 76
De Santiago, Grissel Trujillo 62
Decker, Claire 3
Deo, Shravani 53
Devaull, Christian 93
Dickie, Julia 81
Ding, Zhenya 86
Discher, Dennis E. 32
Diwakar, Prasoon K. 73
Dominick, Gregory 83
Dooling, Lawrence J. 32
Durmusoglu, Deniz 27
Elliott, Mikaya M. 73
Emerson, Samuel 76
Erickson, Patrick 26
Farag, Veronica 26
Faria, Gabriela 70
Feroze, Abdullah 76
Ferreiro, Aura 1
Fleischman, Angela G. 74
Foster, David G. 5, 7
Gadalla, Dina 44
Garana, Belinda 75
Gilchrist, Malcolm Lane 92
Graham, Nicholas 75
Gwinner, Wilfried 9
Haddad, Katherine 64
Haley, Mercedes 79
Haller, Kristen I. 73
Han, Namshik 74
Haney, Bobby 40
Hanley, Luke 25
Harrison, Roger 70
Hatlen, Rosalyn 90
Hatzikirou, Haralampos 9
Hayes, Brandon H. 32
Heldt, Caryn 3
Henson, Michael 25
Hoops, Jordan 73
Hunt, J Porter 89
Hunt, Kristopher A. 25
Huseby, Eric 78
Hwang, Woo Chang 74
Ibrahim, Dauda 66
Ikhapoh, Izuagie 39
Islam, Sumaiya 79
Jablonski, James 43
Jang, Soojin 47
Jariwala, Soham 83
Jayaraman, Anjana 19
Jensen, Dane 24

Jeppson, Mary 12, 17
Jeruss, Jacqueline 93
Jha, Prateek 11
Johnson, Patrick A. 81
Joshi, Pratik 3
Judson-Torres, Robert 96
Jung, Seongwon 64
Jyoti, Sharmin Sultana 81
Kadu, Siddhant 32
Kang, Junhyuk 19
Kannan, Kavya 76
Karakitsios, Spyros 59
Karakoltzidis, Achilleas 59
Karch, Clément G. 70
Kardar, Mehran 78
Karki, Prashant 33
Keene, C. 76
Kennedy, Maeve 44
Khan, Anowar H. 88
Khan, Imran 64
Kieslich, Christopher 35, 52
Kim, Jae Sun 47
Kim, Mingin 63
Kirby, Brian J. 19
Kis, Zoltán 66
Kondapaneni, Raghu Vamsi 68
Kontoravdi, Cleo 66
Krishnamurthy, Ravi 41
Kumta, Prashant 41
Kwon, Young Jik 74, 82
Lane, Nancy E. 64
Larson, Shaylee R. 12
Latorre, Rocco 24
Lee, Rebecca 74
Leggett, Susan E. 30
Lei, Cheng Wai “winne” 74
Lei, Pedro 39
Leland, Paige 12
Lesi, Adeyinka 34
Lewis, Parker 24
Lewis, Randy S. 14
Li, Wei 86
Li, Yueming 92
Liu, Song 43
Lott, David 44
Lugin, Margaret 74
Lym, Emily 17
Manwaring, Kyle 14
Martin, Elizabeth C. 88
Martinez, Sophia 30
Martinez-Paniagua, Melisa 53
McDonald, Karen A. 64
McGill, S. Lee 25
Mehrotra, Pihu 39, 43
Melton, Samuel 78
Melvin, Adam 88
Menon, Prashant 53
Mihalas, Anca 76
Miller, Nicole 73
Mir, Mahrukh 23
Mittal, Shachi 56
Mizerak, Ben 41
Mohammadi, Moein 81
Mohanty, Swomitra 12, 17
Moon, Tae Seok 1
Moustafa, Tarek 96
Nandi, Somen 64
Nelson, Celeste M. 30
Orbach, Sophia 93
Orman, Mehmet 33
Pantazes, Robert 79
Papathanasiou, Maria 66
Parekkadan, Biju 26
Park, Chul Min 47
Park, James 76
Patel, Anoop 76
Pease, Leonard 58
Peng, Ching-An 80
Peng, Ian 80
Perry, Sarah 3
Personius, Kirkwood 43
Pinkerton, Nathalie M. 24
Pojman, John 88
Poklar, Miranda 41
Polanco, Edward 96
Pollard, Rachel 24
Poynter, Felicia 56
Quesada, Braulio Ortega 88
Rajabian, Nika 39, 43
Rajagopalan, Padmavathy 90
Ramakrishnan, Subramanian 40
Rao, Shreyas 68
Redus, Lauren 7
Reichel, Kate 53
Reppas, Andreas I. 9
Rong, Na 39
Roth, Charles 26
Rottinghaus, Austin 1
Roychowdhury, Swarnima 26
Roysam, Badrinath 53
Rubinstein, Michael 4
Ruiz, José Luis Martinez 27
Rumschitzki, David 34
Ryu, Hyung Chul 47

Safaei, Zahra....36
Salsbury, Timothy....58
Sant, Himanshu....12
Sarigiannis, Dimosthenis....59
Sarkis, Miriam....66
Sathyavageeswaran, Arvind....3
Savvopoulos, Symeon....9
Scheffner, Irina....9
Schmidt, Brian....24
Seldeen, Kenneth....39
Sensenbach, Shayne....33
Seong, Hasoo....47
Shah, Nilay....66
Shah, Vidit....56
Shahini, Shahryar....43
Shea, Lonnie....93
Shekhar, Karthik....54
Soltani, Mehran....89
Song, Hyeju....35, 52
Song, Hyun-Seob....47
Suminski, Richard....83
Tarar, Ammar Ahmad....80
Thiyagarajan, Ramkumar....39
Thompson, Gary....36
Thone, Melissa N.....82
Thurber, Greg....49
Tirumkudulu, Mahesh S.....23
Todd, Lucy....85
Toley, Bhushan....15, 28
Troen, Bruce....39
Tseropoulos, Georgios....39
Turkarslan, Serdar....76
Udin, Susan....43
Ugaz, Victor....63
Vanbeek, Matthew....77
Varadarajan, Navin....53
Varela-Soler, Yadiel....26
Vitt, Madison....81
Vlachokostas, Alex....58
Wagner, Norman J.....83
Wall, Gerard....81
Walton, S. Patrick....31
Wang, Hua....72
Wang, Jianmin....43
Warren, Rachel....68
Wawrousek, Karen....81
Wiegand, Connor....41
Woodward, Alexis....70
Wu, Kwan-Ling....53
Wu, Wei-Ju....76
Yellam, Kiranmai....11
Yin, Rose....78
Yung, Yeni....25
Zangle, Thomas....96
Zeng, Xianci....3
Zhang, Yali....43
Zhou, Sophia....88
Zhou, Tingtao....22

AIChE
120 Wall Street, FL 23
New York, NY 10005-4020

ISBN 978-1-7138-7915-2